The Delaware Bay at War!
The Coastal Defenses of the Delaware Bay during World War Two

By Terrance McGovern

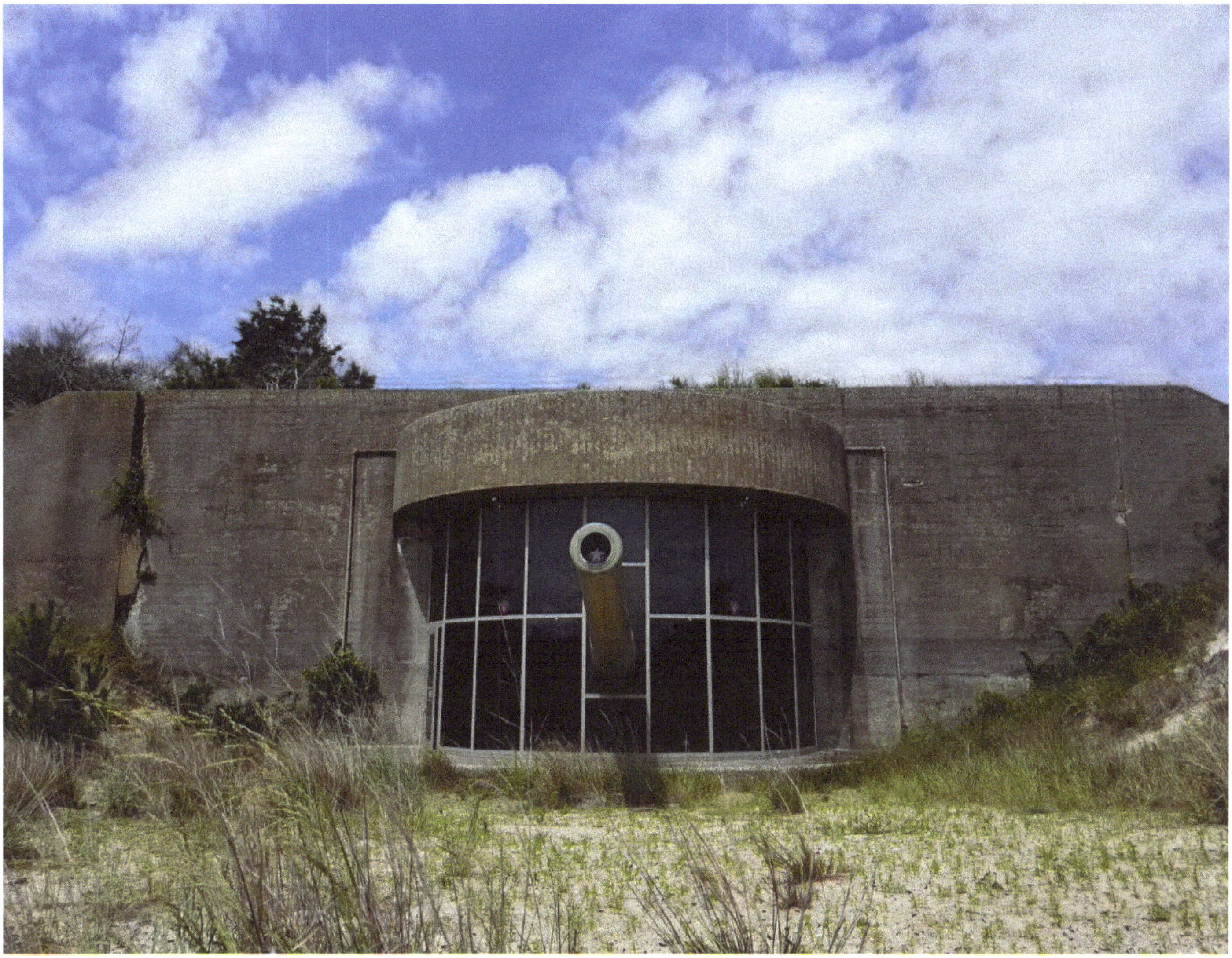

Battery 519, Fort Miles, Cape Henlopen State Park, Delaware with 12-inch barbette gun installed in the south gun casemate. (McGovern Collection)

A Division of McGovern Publishing

COPYRIGHT © 2022 by Terrance McGovern

PHOTOGRAPHS ARE IDENTIFIED IN THE CAPTIONS AS TO SOURCE. ALL PHOTOGRAPHS ARE COPYRIGHTED BY THE SOURCE, UNLESS THE SOURCE IS THE UNITED STATES GOVERNMENT, IN WHICH CASE PHOTOGRAPHS ARE IN THE PUBLIC DOMAIN.
ALL RIGHTS RESERVED. NO PART OF THIS BOOK MAY BE USED OR REPRODUCED WITHOUT WRITTEN PERMISSION OF THE AUTHOR.

PLEASE DIRECT ANY COMMENTS OR CORRECTIONS TO THE AUTHOR AT tcmcgovern@att.net

IBSN 978-1-7323916-4-2
LIBRARY OF CONGRESS CATALOG CARD NUMBER 2022942757

First Edition: August 2022
Layout by Stephen Dent (sfdent@dircon.co.uk)

Front Cover Photo: 12-inch Gun Model 1917 on Barbette Carriage, Battery Hall, Fort Saulsbury, Delaware (NARA 1941)
Title Cover Photo: Battery 519, Fort Miles with replacement 12-inch barbette gun (McGovern Collection)
Rear Cover Photos: (1) The newly commissioned USAMP Lieutenant William J. Sylvester *(MP-5)* (DNREC 1942); (2) Battery Smith's 16-inch MkII gun (DNREC 1945); (3) A 155mm GPF battery ammunition supply for conducting firing practice at Fort Miles, DE in November 1941 (DNREC 1941); (4) The sinking of the tanker SS Resor by U-578 on February 27, 1942 (RG80 NARA 1942); (5) The 6-inch M1903A2 gun and shielded barbette carriage M1 of Battery Hunter (#222) at Fort Miles, DE (DNREC 1944); (6) US Army Mine Planter *Schofield* prepares to drop a M1 mine and anchor (NARA 1942); (7) Construction at Fire Control Tower No. 3 at Dewey Beach, DE on January 23, 1942 (RG77 NARA 1942).

Authors Bio: Terrance McGovern has authored eight books and numerous articles on fortifications, four of those books being for Osprey's Fortress Series (*American Defenses of Corregidor and Manila Bay 1898-1945*; *Defenses of Pearl Harbor and Oahu 1907-50*; *American Coastal Defenses 1885-1950*; *Defenses of Bermuda 1612-1995*). He has also published 12 books on coast defense and fortifications through Redoubt Press or CDSG Press. Terry was Chairman of the US-based Coast Defense Study Group and continues to be a long-time officer. He has also been the editor of the Fortress Study Group annual journal, *FORT*. He is a director of the International Fortress Council and the Council on America's Military Past. He can be contacted at tcmcgovern@att.net.

Publisher Info: McGovern Publishing, operates through the Three Sisters Press and the Redoubt Press, publishes books of historical interest, especially seacoast fortifications. Under the Redoubt Press label, the firm has published *The American Defences of the Panama Canal* by Terrance McGovern, *The Concrete Battleship - Fort Drum, El Fraile Island, Manila Bay*, by Francis J. Allen, *A Legacy in Brick and Stone*, by John R. Weaver II, *Pacific Ramparts – a History of Corregidor & the Harbor Defenses of Manila & Subic Bay* by Glen M. Williford, and *Pacific Fortress – the Harbor Defenses of Oahu, Hawaii* by Glen M. Williford. Under the Three Sisters Press label, the firm has published *The Chesapeake Bay at War! – The Coastal Defenses of Chesapeake Bay During World War Two* by Terrance McGovern, *Seacoast Cannon Coloring Book* by Brian Chin, and *The Delaware Bay at War! – The Coastal Defenses of Delaware Bay During World War Two* by Terrance McGovern. McGovern Publishing is interested in new titles, especially those dealing with fortifications. Please contact Terry McGovern at 703/538-5403 or at tcmcgovern@att.net if you have a title that you are seeking to have published. Visit www.mcgovernpublishing.com

Redoubt Press
A Division of McGovern Publishing
1700 Oak Lane
McLean, Virginia 22101 USA
tcmcgovern@att.net

A Division of McGovern Publishing
1700 Oak Lane
McLean, Virginia 22101 USA
tcmcgovern@att.net

Table of Contents

Dedication – Gary Wray   4

1. Introduction   5
2. Background   6
3. Fort Saulsbury and Focus on Defending the Bay   8
4. Preparing for World War Two   10
5. Coastal Defenses Move to the Delaware Capes   16
6. War Comes to the Delaware Bay   25
7. German U-boats Patrol the Approaches to the Delaware Bay   35
8. Permanent Coast Artillery Batteries Come to the Delaware Bay   38
9. Manning the Coast Defenses of the Delaware Bay   46
10. Drawdown of Delaware Bay Defenses to Support Overseas Operations   59
11. Delaware Bay Defenses after World War Two   66
12. Delaware Bay's World War Two Defenses Today   69

Appendix –
- Location Maps   90
- Fort Miles Historical Association   94
- Coast Defense Study Group   96
- McGovern Publishing   98

Firing Practice on 90mm Mobile AA Gun at Fort Miles (DENG 1955)

Gary Wray, Co-founder of the Fort Miles Historical Association

Dr. Gary David Wray, 78, of Lewes, Delaware passed away on Thursday, Feb. 3, 2022, after a very short battle with cancer. Gary was a CDSG member and served on the 2015 Annual Conference committee to the Defenses of Delaware Bay. He was the driving force behind the preservation and interpretation of the WWII Fort Miles. Gary was schedule to write a Forward for this book, but unfortunately, we have a Dedication instead. Gary strongly supported the publication of this book.

Gary was a proud West Virginian and a 1965 graduate of Morris Harvey College (now University of Charleston) in Charleston, where he majored in his lifelong passion – history – with a minor in another lifelong passion, education. His entire life, Gary would be a thorough source of historical facts – he was an amazing storyteller. Gary came to Delaware in 1966 from West Virginia and continued his career in education in Delaware as a teacher and administrator in the Caesar Rodney, Milford and Cape Henlopen districts. During his 30+ years in the Cape Henlopen School District, Gary was a teacher at Cape Henlopen High School in the 1970s, principal at Lewes Junior High School and Lewes Middle School in the 1980s and 1990s, and director of secondary education from 1984-93, retiring as a Cape Henlopen School District administrator in 1995. He was elected to the Cape Henlopen School Board in 2005 and served from 2005-10; he was vice president of the board for one year and president for three years. During his tenure, he actively worked to build the new Cape Henlopen High School that opened in 2010, and Gary helped to create the Cape Henlopen Educational Foundation (CHEF). Gary also was familiar with the charter school system, as he worked to create the charter for Sussex Academy and served on the board of directors as a founding member until he resigned from his Sussex Academy board seat in 2001. Gary also was a university professor of history and many other subjects, both in undergraduate and graduate programs, in many colleges and universities throughout Delaware and Maryland, primarily at Wilmington University for many decades.

From the moment Gary moved to Delaware in 1966, his passion for history was involved in all aspects of his life in Delaware, first working on improving Fort Delaware and later in the 1990s shifting his attention to Fort Miles in Lewes. In August 2003, Gary co-founded the Fort Miles Historical Association (FMHA) with Bob Frederick, David Main, and the late Delaware State Parks Historian Lee Jennings, while standing under FC Tower 3 at Dewey Beach. In 2005, Gary and Lee wrote the book on Fort Miles, published by Arcadia Publishing. During his 17-year tenure with FMHA, Gary served as the organization's president, combining his love of history, particularly military history, with his extraordinary vision and dedication to preserving history and serving his community, to lead FMHA to significant growth and the formation of the Fort Miles Museum in 2016. Gary's goal was to work with others to make the Fort Miles Museum the best World War II Museum inside a WWII facility in the country. John Roberts, an association board member and head of the working group called the "Bunker Busters", worked closely with Gary. "Without him and the other founders, Fort Miles would be a muddy hole in the ground," he said. "His legacy was as an unstoppable force to be reckoned with. The main part of our job was to make sure his passion was directed," Roberts said. Gary was instrumental in saving one of the USS *Missouri*'s 16-inch barrels by bring it to Cape Henlopen State Park. Gary took on a mission to preserve the railroad history of Lewes as one of the founders of the Lewes Junction Railroad & Bridge Association (LJRBA). Civic duty and saving historical artifacts were lifelong passions for Gary Wray, and the members of FMHA and LJRBA are continuing those efforts.

Gary was elected to the oldest Delaware historical society, the Delaware Historical Society Board of Trustees, and served two terms. In 2016, Gary was chosen for the Delaware Maritime Hall of Fame when the Overfalls Foundation held its 10th annual Delaware Maritime Hall of Fame induction. Gary's name is engraved on the Hall of Fame monument, which sits along the canal in Lewes adjacent to the Lightship Overfalls. Fort Miles always had a very special place in Gary's heart.

The defense of America's seacoast has been one of the key concerns since the earliest years of the Republic. American coast defense steadily evolved through the age of muzzle loading cannon, ever larger breech loading weapons, and finally to the culmination in large, long range guns capable of targeting the largest and most heavily armed warships of their age. By the end of World War Two, the United States had some of the strongest coastal defenses in the world.

Delaware Bay is the estuary outlet of the Delaware River on the northeast seaboard of the United States. Approximately 782 square miles (2,030 km²) in area, the shores of the bay are largely composed of salt marshes and mudflats, with only small communities inhabiting the shore of the lower bay. Besides the Delaware, it is fed by numerous smaller rivers and streams, including (from north to south) the Christina River, Appoquinimink River, Leipsic River, Smyrna River, St. Jones River, Mispillion River, Broadkill River and Murderkill Rivers on the Delaware side, and the Salem River, Cohansey River, and Maurice Rivers on the New Jersey side. The bay is bordered inland by the States of New Jersey and Delaware, and the Delaware Capes, Cape Henlopen and Cape May, on the Atlantic.[1]

The Bay has always been a conduit to ports along the Delaware River – Philadelphia, Wilmington, Trenton, Chester, Camden, and other smaller cities - home of many heavy industries such as chemicals, ship building, steel, textiles, and oil refining. Given the importance of the military-industrial complex along the banks of the Delaware River, including the large Philadelphia Naval Shipyard and explosive factories of the E. I. du Pont de Nemours & Company, the defense of the Delaware River had a prominent role in America's military planning from the American Revolution to the First World War. A barrier of three forts (Fort Mott, Fort Delaware, and Fort Du Pont) defended the river at the end of the nineteenth century. In the years leading up to World War Two, the coastal defenses protecting Delaware River had declined dramatically leaving them mostly on caretaker status with

Figure 1. Fort Delaware on Pea Patch Island guards the shipping channel to the Delaware River. Battery Torbert (3-12DC) built within while Battery Hentig (2-3P) and Battery Dodd (2-4.7P) are located outside the 3rd System fort. Battery Hentig remained active during WW2 until the guns were transferred to Battery 5 at Fort Miles in May 1942. (McGovern Collection 2015)

only small units of the US Army's Coast Artillery Corp available to man these defenses. The prospect of a new world war and the advances in military technology resulted in new investment in coastal defenses starting in the late 1930s.

The defenses of the Delaware River shifted to the Atlantic seacoast and entrance to the Bay after World War One. During World War Two, the primary military reservation was Fort Miles on Cape Henlopen, near Lewes, Delaware, as well as its sub-post on Cape May in New Jersey. The outbreak of war saw temporary, mobile coast artillery, such 155mm GPF guns and 8-inch railway guns, as well as 3-inch rapid fire guns deployed to the Capes to provide coverage as permanent coast defenses were constructed. These permanent coastal defenses featured long-range sixteen-inch and twelve-inch guns, as well as 6-inch secondary batteries, and smaller rapid-fire seacoast and anti-aircraft batteries. Additional, controlled submarine mines were placed at the entrance to the Bay, along with submarine nets and torpedo boat booms within the Bay. Supporting the US Army's defenses, was the US Navy with its inshore patrol craft, aircraft, and blimps, while the US Army Air Corp provided aircraft and the US Ground Forces provided mobile combat teams (protection

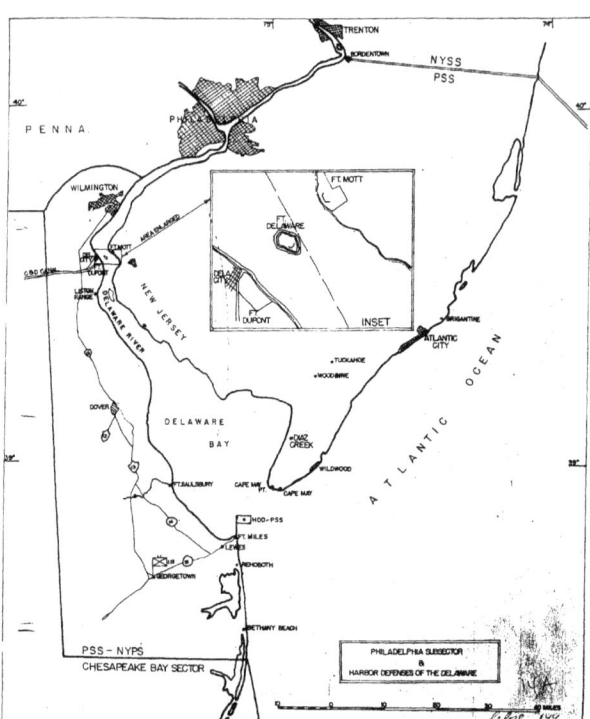

Figure 3. The Philadelphia Subsector and the Harbor Defenses of the Delaware during World War Two. (NY-Phil History 1945)

against enemy landings). While enemy battleships never threaten the Bay's defenses, German U-Boats were active in the waters off Delaware Bay causing residents to see and hear the destruction of commercial shipping.

This program of modern coast artillery batteries and other seacoast defenses would take four years to construct and require millions of dollars before reaching their reached their apex during the middle of World War Two, yet by 1948 be almost completely gone. As part of this modernization program many silo-shaped concrete fire-control towers were constructed along the New Jersey coast as far north as Wildwood and down the Delaware coast to Ocean City, Maryland to allow the coast artillery to target enemy warships. Today, these surviving fire-control towers and associated concrete bunkers invoke speculation by visitors on what role did these structures serve. It is hoped that this book will help answer those questions.

Background

Over the span of nearly two centuries, the nature of the defenses on the Delaware River and Bay changed as advances were made in the technology of armament. During the colonial period, blockhouses and log forts were built to provide a semblance of local protection. One of

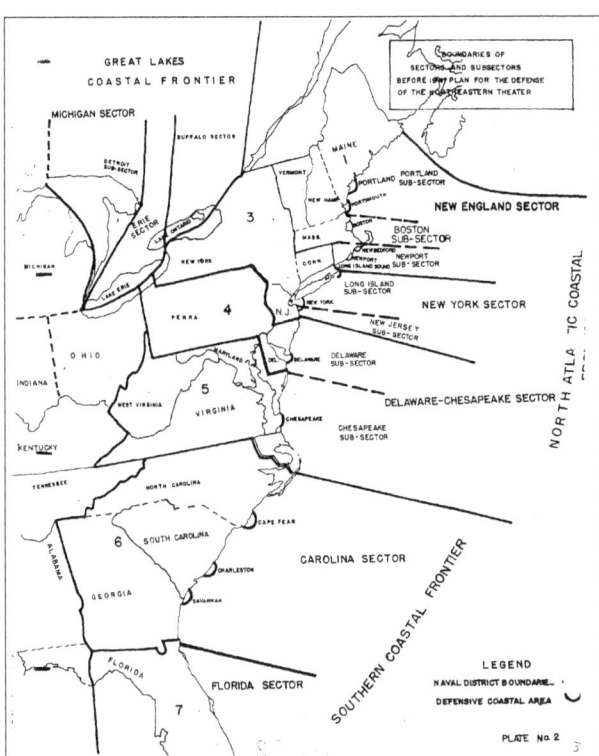

Figure 2. Naval District Boundaries on the eastern seaboard of the US (NY-Phil History 1945)

Figure 3B: Firing Practice with the 12-inch mortars at Batteries Rodney-Best at Fort DuPont, Delaware in May 1918 (NARA 1918)

the earliest efforts to provide defenses in what would become the area of Philadelphia was Fort Nassau, erected by Dutch colonists in the 1620s at Gloucester Point near Red Bank. Other works were built as the number of colonists and their settlements along the Delaware River increased. Rivalries between the European nations, and the wars they fought to gain control of more territory in the new world, prompted the construction of additional defenses. By the middle of the eighteenth century, during the struggles between the French and British for control of North America, the first efforts to defend Philadelphia from naval attack were made when a battery of guns were mounted on Society Hill in that city c.1750.[2]

From that modest beginning came a series of works that reflected the evolution of coastal defense: from the ill-conceived works begun by the British on Mud Island in the early 1770s (today known as Fort Mifflin) to the ill-starred Fort Delaware of 1815-1831 and the much larger third system fort that succeeded it; to Forts DuPont and Mott erected at the turn of the century, which gave way in turn to the powerful batteries emplaced further down Delaware Bay at Fort Saulsbury in the post-World War One years. These were to be supplanted in World War Two by the massive casemated long-range batteries, and their supporting batteries, on Cape Henlopen at the entrance to Delaware Bay.

The coastal defenses of Delaware River and its principal cities throughout the American Revolution, the American Civil War, and World War One are important histories but are outside the scope of this work. These older defenses will play only a minor role during World War Two as the active defenses will shift away from the River itself to focus on the Bay and its entrance. The 1920s and 1930s had seen a sharp decline in manning of these fortifications and most were on caretaking status or had their armament removed for use in Europe in the World War One. Most of these fortifications were built as part of the Endicott-Taft Program to modernize America's seacoast defenses after years of decline since the American Civil War. In 1886, the Board on Fortifications or Other Defenses, more familiarly known as the Endicott Board after its president, US Secretary of War William C. Endicott, recommended the construction of detached batteries behind earthen parapets surmounting and protecting concrete magazines, bomb proofs, and storerooms. Armament was to consist of breech loading steel rifles and mortars in various sizes, mounted on disappearing carriages, barbette carriages, and turrets, and supported by minefields and torpedo boats.

As a result of the Endicott-Taft program, between 1899 and 1924 the the armament shown in Table I was installed at Fort DuPont, Fort Delaware, and Fort Mott (in New Jersey) to form a three-fort barrier across the River at Pea Patch Island, and during the post-World War One period, Fort Saulsbury, was added at Slaughter Beach in Delaware.[3]

By the beginning of World War One, much of this armament was obsolete or soon to be removed for the war effort and the rest at best obsolescent compared to the newer big gun dreadnoughts. Advances in naval armament had made 15-inch and 16-inch gun battleships formidable antagonists for any fixed fortifications. The future sites of Fort Miles on Cape Henlopen and Cape May's military reservation were used in May 1918 for temporary batteries mounting one 6-inch gun at each point to provide for examination batteries.[4] For the first time, anti-aircraft guns were installed at key industrial factories along the Delaware River as well as at Fort DuPont.[5 6 7]

The advent of coast artillery with long range meant that American military planners could now place batteries near the entrance to Delaware Bay that could reach an enemy warship as the entered the bay thus denying their access to both the River and Bay. This would also allow for the three-fort defenses to become secondary and reduced both the number of batteries and manpower (thus lowering costs as well). This transition from the River to the Bay in the post-World War One period resulted in construction of Fort Saulsbury and its four 12-inch guns on long-range carriages. Fort Saulsbury represented the first step in defending the entire Bay in the inter-war period.

Table 1 Delaware Forts 1899-1924

FORT MOTT, Finns Point, N.J. established 1900 now a state park

Battery Arnold	*Battery Harker*	*Battery Gregg*	*Battery Krayenbuhl*	*Battery Edwards*
3 x 12in on disappearing carriage. In service 1899-1943.	3 x 10in on disappearing carriage. In service 1899-1941. Guns and carriage to Canada, guns still at location in Canada.	2 x 5in on P mount. In service 1901-1910.	2 x 5in on BP mount. In service 1900-1918.	2 x 3in on CM moun.t In service 1902-1920.

FORT DELAWARE, Pea Patch Island. Established 1847 now a state park /MC

Battery Torbert	*Battery Dodd*	*Battery Hentig*	*Battery Alburtis*	*Battery Allen*
3x 12in on disappearing carriage. In service 1901-1942. Guns to Battery Reed, Puerto Rico.	2 x 4.7in on A mount. In service 1899-1918.	2 x 3in on P mounts. In service 1901-1942.	2 x 3in on MP mounts. In service 1901-1920.	2 x 3in on MP mounts. In service 1901-1920.

FORT DUPONT, Delaware City, Del. Established 1896 now part owned by a redevelopment corporation, state agencies & park /MD, MC

Battery Rodney	*Battery Best*	*Battery Read*	*Battery Gibson*	*Battery Ritchie*
8 x 12in mortars. In service 1900-1941.	8 x 12in mortars. In service 1900-1941.	2 x 12in on barbette carriage. In service 1899-1918. Emplacements split. Battery Gibson located between.	2 x 8in on disappearing carriage. In service 1899-1917. One structure with Battery Read.	2 x 5in on P mount. In service 1900-1917. Destroyed.
Battery Elder	*Battery Elder II*			
2 x 3in on P mount. In service 1904-1922. Relocated 1922 to Delaware Beach	2 x 3in on P mount. In service 1942-1942. Liston front range light			

FORT SAULSBURY, Slaughter Beach, Del. Established 1917. Now in private ownership, Wildlife Refuge.

Battery Hall	*Battery Haslet*
2 x 12in on long range barbette carriage. In service 1924-1945.	2 x 12in on long range barbette carriage. In service 1924-1942. Guns to Ft. Miles.

Fort Saulsbury and Focus on Defending the Bay

In 1916, little time was lost in selecting a site for the two newly authorized long-range 12-inch batteries. As soon as the design for a long-range barbette carriage was developed, the process of procuring sites began. The location for the two batteries on Delaware Bay was on the Delaware shore near where the Mispillion River empties into the bay, some six miles east of Milford. Negotiations began in February 1917 for farm and swamp land amounting to just over one hundred and fifty-one acres by purchase and condemnation. The onset of American involvement in World War One hastened these efforts, which were concluded in June, and the engineering work began.[8] Preparation of the site was advanced sufficiently by August to commence the two batteries. The batteries were built concurrently, and were prosecuted with vigor during the war. Early in 1918, while the magazine traverses were still under construction, the four 12-inch M1895M1 guns were shipped to the battery sites by barge and mounted on their long-range all-around fire M1917 barbette carriages. Once the guns were mounted, a small manning detachment of coast artillery troops was posted there from Fort DuPont. With the signing of the armistice, the troops were withdrawn and the pace of construction slowed. In December 1920, some forty months after the batteries were begun, they were finally completed. On 27 December, 1920, the batteries were transferred to the custody of the commanding officer, Coast Defenses of the Delaware.[9]

Fort Saulsbury was a modest post. It consisted of the two batteries, a handful of temporary wooden buildings for officers' quarters, quarters for the engineer, and two storehouses for the ordnance and engineer stores. Initially, the small wartime manning detachment was housed inside the massive magazines. Permanent quarters for a manning detachment would not be provided for nearly a decade.[10] The earth covered combination magazine and service structures between each pair of guns looked like massive flat-topped hills or dunes rising above the marsh. The main body of the reinforced concrete structure was 125 feet long with a 33-foot extension at each end, while the centers of the gun emplacements were some 420 feet apart. The width of the structure was about 125 feet. Because of the somewhat isolated location, quarters were built into the battery structures. Along the rear of the building, a series of rooms opened onto the battery's sixteen-foot-wide rear corridor. These bombproof rooms were equipped with fireplaces for warmth in the winter and helped to keep the interior spaces of the structure dry. Across this corridor were the powder magazines, shell rooms, electric power generator room, storerooms, and the plotting rooms. Steel spiral stairways on either side of the emplacement entrances led up to the battery commander's

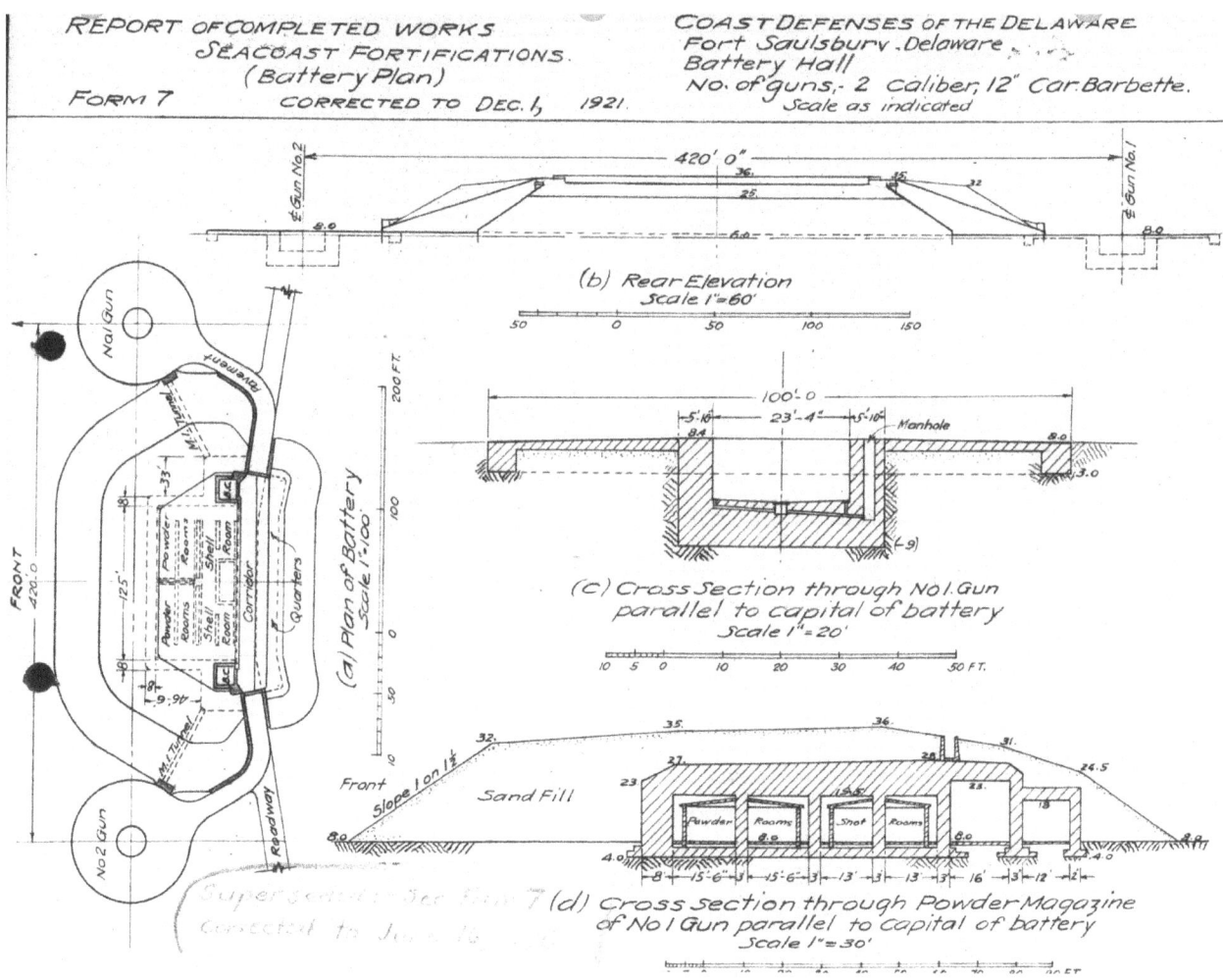

Figure 4. RCW Form 7 of Battery Hall, Fort Saulsbury, DE. Batteries Hall and Haslet were identical in design and armament. (RG77 NARA 1921)

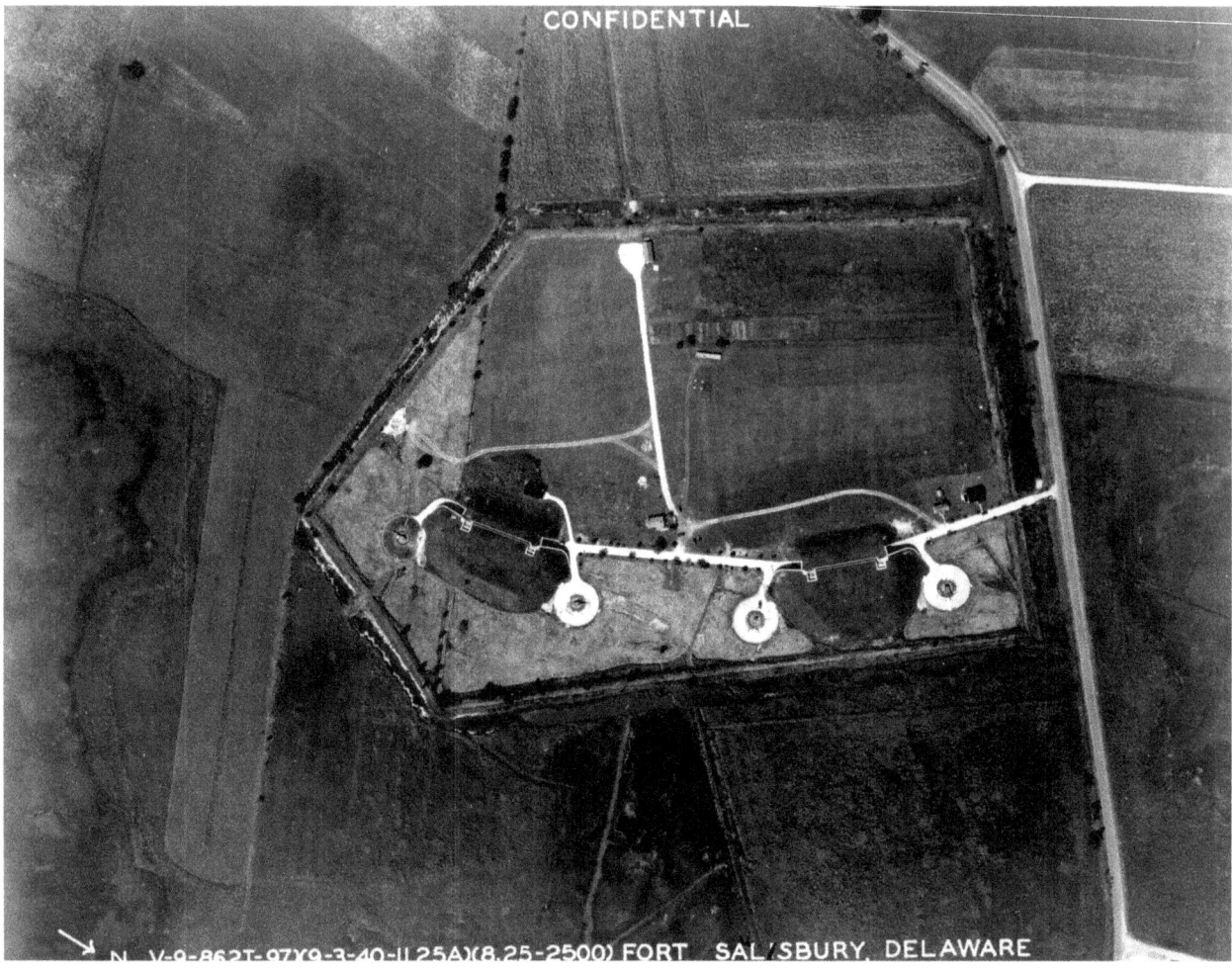

Figure 5. Aerial view of Fort Saulsbury at Slaughter Beach, DE on September 9, 1940. (NARA 1940)

stations on top of the traverse.[11] An overhead monorail trolley system helped move projectiles in the shell rooms. Powder was run from the magazines to a truck corridor at either end of the magazine spaces where the powder charges were placed on hand trucks and moved to the guns. The projectiles were similarly moved from the shell rooms to the truck corridors and thence to the corridor where they were lowered from the trolley to projectile trucks and pushed to the guns through the rear corridor. [12]

At some point following completion of the battery, a concrete fire control station was built near the bay front in the rear of the Mispillion Light House. This station and the battery commander's stations atop the battery traverses, some 35 feet above the battery rear corridors provided the fire control system for Batteries Hall and Haslet until the early 1940s, when additional base end stations were constructed along the shoreline of the bay. The circular loading platforms for the 12-inch guns measured one hundred feet across, while the gun block in the center of the emplacement had a diameter of thirty-five feet. From the air, the circular platform gave the appearance of a giant bullseye; soon a potentially easy mark for enemy aircraft. During the period that followed the batteries' completion and the beginning of World War Two, only a few settling rounds were fired from the two batteries.[13] Fort Saulsbury would serve as Delaware Bay's "modern" defenses from end of World War One into the beginning of World War Two. During much of this time the fort was on caretaker status and unmanned. Only the coming of World War Two would see the fort's two batteries become activity and ready to defend the Bay.[14]

Preparing for World War Two

The lack of activity and absence of a coast artillery garrison for the forts on the Delaware River after World War One did not mean the US War Department placed a small value on the defenses of the Delaware. By 1928, the age of the armament in the harbor defenses, combined

Figure 6. 12-inch M1895M1 gun on M1917 long-range barbette carriage (with 75mm M1916 sub-caliber practice gun mounted on top of the 12-inch gun) at Battery Hall, Fort Saulsbury. (NARA 1941)

with the isolation and faulty tactical location of the Delaware forts, were recognized by the War Department. Brig. Gen. Stuart Hentzelman, commanding general of the Second Coast Artillery District, noted that "projects under consideration contemplate placing the defenses at one or both of the Delaware Capes (May or Henlopen), which will no doubt be realized when funds become available and will, if practicable, include the use of Fort Saulsbury." When the development of harbor defense projects was ordered by the War Department on 12 August, 1931, Delaware Bay was included.

The basic harbor defense project for Delaware Bay was submitted on 9 November, 1932, approved 21 March, 1933, by a board of officers, and by the U.S. Secretary of War on 11 May of that year. The Harbor Defenses of the Delaware were defined as the Atlantic coastline from the north side of Hereford Inlet, N.J., past the entrance to Delaware Bay to the south side of Rehoboth Bay, Del. In addition, the harbor defenses were responsible for the security of the cities of Philadelphia, Chester, Newcastle, and Wilmington on the Delaware River, and for the protection of the Delaware Bay entrance to the Chesapeake and Delaware Canal.[15]

The harbor defense mission stated was:

- To insure freedom of movement of [U.S.] naval vessels in and out of Delaware Bay.
- To afford an anchorage in Delaware Bay and River secure from naval gunfire.
- To deny to the enemy access to Delaware Bay and River.
- To support the defense against landing attacks within the range of harbor defense weapons.

It should be noted that the U.S. Army's Coast Artillery Corps did not attempt to prevent landings at all points along the coast, but only to protect vital harbors. The defense of the remainder of the coastline was assigned to units of the mobile army.

The means of carrying out this mission were pitifully small in 1933. The only guns capable of even attempting to cover the bay's entrance were the four long range

12-inch guns at Fort Saulsbury. Although their field of fire covered the interior waters of the bay very well, they could barely reach the bay's entrance. To defend the lower reaches of the bay and effectively cover the seaward approaches, additional armament would be required.

The idea that the Delaware Capes should be sites for defenses was not a new one. As early as the 1830s, the War Department had considered construction of defensive works at Cape Henlopen, and on 20 January, 1836, Col. Charles Gratiot, then chief engineer, had submitted to the Committee on Military Affairs of the House of Representatives an estimate of $150,000 to build a second-class fort on Cape Henlopen to cover the harbor of refuge inside the cape and the breakwater.[16] Although no works were commenced, the prospect of permanent fortifications for the site had been considered for a century when the basic harbor defense project for Delaware Bay was developed.

In time of war, however, the project recommended that a satisfactory defense could be provided economically by a battery of two 14-inch guns and a battery of two 8-inch guns, all on railway mounts at Cape Henlopen. This armament would also be sufficient to keep enemy warships at a distance and permit the safe sortie of U. S. Navy vessels from the bay. In addition to the railway guns, a battery of 155 mm guns "without concrete platforms" on Cape Henlopen, would, in the mind of the board, be sufficient to cover a controlled minefield at the bay's entrance. Another battery of 155 mm guns on Panama mounts at Cape May would deter vessels of lighter draft from entering the bay through the shallow channels near that cape. Construction of the Panama mounts was not to be carried out until an emergency arose. Pending procurement of the 14-inch guns, the project recommended that a pair of 12- inch railway guns on batignolle carriages should be substituted and that the "necessary parts for modification of the mounts and carriages for the use on a fixed platform be secured and held available."[17]

Should an enemy naval force manage to enter Delaware Bay, it would then come under the fire of Batteries Hall and Haslet at Fort Saulsbury, whose 12-inch guns covered most of the deep waters of the bay as well as the approaches to the Delaware River. These two batteries, however, still lacked the comprehensive fire control system that would enable the batteries to effectively control the interior waters of the bay.

Farther up the bay, the entrance to the Chesapeake and Delaware Canal was covered by two batteries of 3-inch guns. Battery Elder was at Fort DuPont, while Battery Hentig was sited outside Fort Delaware on the south end of Pea Patch Island. In 1933, the approaches to the river and to the canal were covered by a still formidable, although somewhat obsolete, array of armament: the three 12-inch guns of Battery Arnold and the three 10-inch guns of Battery Hacker at Fort Mott, and Battery Torbert's three 12-inch guns at Fort Delaware. At Fort DuPont, Batteries Best and Rodney were also able to fire their eight 12-inch mortars some 14,650 yards down the bay. It was felt, however, that the 12-inch guns of Batteries Arnold and Torbert were sufficient to deny enemy capital ships the

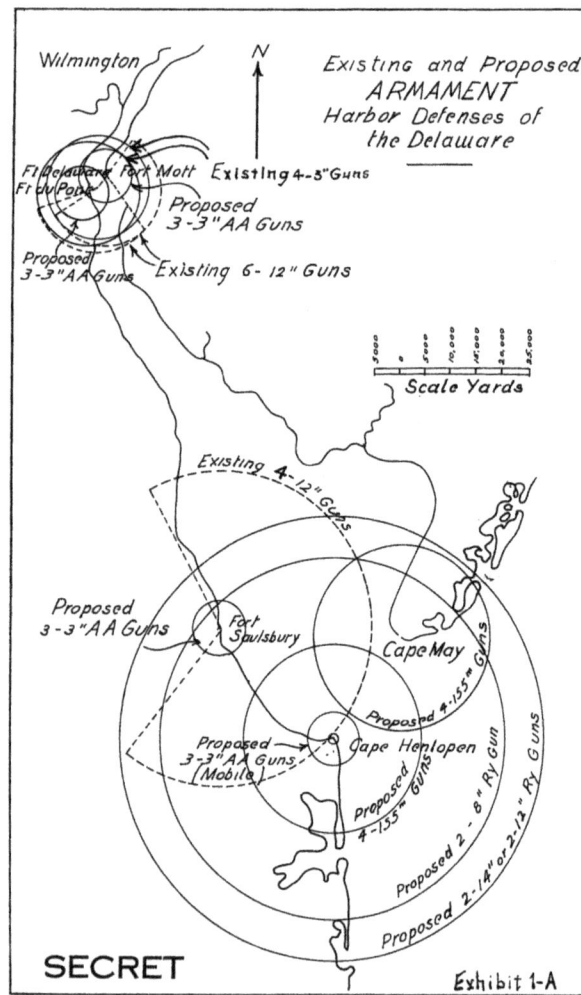

Figure 7. Fields of fire for proposed seacoast batteries in the Harbor Defenses of the Delaware, 1930s. By 1940, the 14-inch railway guns were deleted from the project in favor of a pair of 16-inch gun batteries, and the number of 8-inch railway guns increased from two to eight. (Basic Harbor Defense Project, 1933, Harbor Defense Projects for the Philadelphia-Delaware River Area, RG 177, NARA)

river channel and the armament of Batteries Best, Rodney, and Harker were recommended for elimination from the harbor defense project.[18]

In terms of searchlights in service in the harbor defenses, there was only one, at Fort DuPont. There were, however, several more in storage. Their disposition was to be determined at a later date, but sites for their emplacement further down the bay were not to be purchased except in time of war.

The underwater defense project of Delaware Bay was approved by a joint army-navy board on 12 January, 1931. It called for a controlled minefield of 25 groups laid across the main channel entrance to the bay. In addition, an antisubmarine net was to be placed across the entrance to the harbor of refuge inside Cape Henlopen. The materiel and equipment necessary to implement this project was, however, still being procured, and was far from ready for service.[19] The Navy would be called upon to provide patrol craft as part of the Inshore Patrol Force, based at Cape May, and seaplane patrols of the approaches to Delaware Bay, based out of the Naval Air Station, Cape May.

Another deficiency lay in the antiaircraft gun defenses. There were in 1933 no antiaircraft guns installed in the harbor defenses. The basic harbor defense project indicated that twelve guns would be required. Three batteries, each consisting of three 3-inch AA guns on fixed carriages, were to be provided for Forts Mott, DuPont, and Saulsbury. Three more 3- inch guns on mobile mounts were to be provided at Cape Henlopen. Sixteen antiaircraft searchlights and sound locators were also to be provided, seven between Forts Mott and DuPont, four more at Fort Saulsbury, and another five at Cape Henlopen. Provision was also to be made for antiaircraft machine guns to "furnish protection for ... harbor defense installations from attack by low flying planes."[20]

During the 1930s, the basic project was fine-tuned and the various annexes prepared by the harbor defense staffs specified the various details of the bay's defense plans. By this time the increased reliance upon railway guns in lieu of fixed batteries that had been prevalent following World War One was ending, and railway artillery was deleted from many of the harbor defense projects. On 20 April, 1934, the 14-inch railway guns at Cape Henlopen were deleted from the project and a battery of two 16-inch guns in bombproof casemates was substituted. This casemated gun battery was one of seven similar batteries approved in 1934 for the defense of the nation's principal seaports, using 16-inch MkII naval guns on various modifications of the army M1919 carriages. While construction of at least three of these powerful casemated batteries was initiated, the battery slated for Cape Henlopen was not commenced until the latter months of 1941.

The men for the various elements of the harbor defenses were to be supplied from the regular army and National Guard. On 9 July, 1936, the HQ and HQ detachment, 261st C.A. Bn., Delaware National Guard, was organized, giving the battalion a headquarters unit and two firing batteries. A third battery, Battery C was organized in 1940, at Laurel.[21] Most of the activity of the National Guard troops was confined to their annual summer training. Although Fort DuPont had been utilized between the two world wars for training National Guardsmen, by 1940, this training usually took place at the National Guard training center at Bethany Beach or at nearby Rehoboth Beach, south of Lewes.

In April 1938 the U.S. Army held a training camp at Cape Henlopen for men from the 52nd Coast Artillery, which was based at Fort Hancock, N.J. The 52nd C.A moved over the rail system its equipment from Sandy Hook, NJ to Cape Henlopen where special rail lines had been installed to allow the unit to conduct practice firing in defense of Delaware Bay. This training exercise involved both four 12-inch mortars (12-inch Mortar M1890M1 on M1918 Carriage (Railway)) and four 8-inch rifles (8-inch Gun M1888MIA1 Barbette carriage M1918 on railway car M1918MI), set up about 2,000 yards from the water's edge and behind the Great Dune. The dune served as a screen over which the practicing artillerymen would fire at targets that were 900 to 1,200 yards offshore and towed behind a mine planter that also traveled to Cape Henlopen of the exercise.[22]

In May 1939, Battery E, 7th CA (HD), commanded by Captain Paul A. Harris, consisted of two officers and fifty enlisted men. The battery was the sole active coast artillery unit assigned to the defenses of the Delaware. This organization had been at reduced caretaker strength throughout the 1930s, being posted at Fort DuPont as caretaking for all of the seacoast batteries, as well as installations such as searchlights, fire control stations and equipment, and the submarine mine defenses, in the Harbor Defenses of the Delaware. Battery E supplied three six-man fatigue and maintenance details to Forts Mott, Delaware, and Saulsbury, while the remainder of the caretaking detachment took care of the coast artillery property at Fort DuPont. The generally obsolete armament of the harbor defenses was still considered in commission, although out of service. When war erupted in Europe later in 1939, the battery was recruited up to near peacetime strength.[23]

Orders were received at Fort DuPont to activate the 21st Coast Artillery (HD) Regiment. The 1st Bn., 21st C.A. (HD), had been constituted on January 19, 1940, and activated at Fort DuPont on February 1, 1940.[24] The regimental headquarters and headquarters battery of the 21st C.A. remained the only active elements of the regiment for the next seven months. By August 1940, when a sufficient number of officers and enlisted personnel had been assigned to the harbor defenses, organization of the first of the 21st's firing batteries was authorized. Battery A was activated August 1, 1940, at Fort DuPont.[25] Colonel George Ruhlen, with the Organized Reserves as senior instructor at Omaha, received orders in early autumn, 1940, transferring him to Fort DuPont to assume command of the Harbor Defenses of the Delaware and the 21st C.A.[26]

Colonel Ruhlen found the defenses of Delaware Bay woefully inadequate. The heaviest weapons in the forts could be easily outranged by guns afloat, none of the batteries had overhead cover, and all were totally vulnerable to attack by enemy aircraft. Air defense was nonexistent. The two antiaircraft guns once emplaced atop the 1870s Barbette Battery at Fort DuPont had been shipped to Panama in 1923 and never replaced. The shortage of antiaircraft guns in the nation's arsenal was so

Figure 8. Gunners swabbing out the barrel of an 8-inch M1888M1A1 railway gun during target practice of the 52nd Coast Artillery Regiment at Cape Henlopen, Lewes, DE on May 25th 1938. (NARA 1938)

critical by the summer of 1940, however, that the War Department stated that neither mobile nor fixed antiaircraft guns could be provided for harbor defenses such as those on the Delaware.

The US War Department was well aware of the shortcomings of American coast artillery compared to modern warships. By the early 1920s, the US Army had developed a long-range 16-inch barbette carriage gun as the standard weapon against capital ships. Little work, however, was actually done to improve the harbor defenses and only several of these 16-inch batteries were constructed. When the ultimate fate of the British and French fleets became a matter of concern in 1940, a

Figure 9. Tent encampment and supporting rail cars of the 52nd Coast Artillery Regiment at Cape Henlopen, Lewes, DE on May 25th 1938. In background, are the fish processing plants that would provide strong odors for the troops stationed at Fort Miles during WW2. Now the site of Cape Shores housing community. (NARA 1938)

complete reassessment of harbor defense was undertaken. In March 1940, the Harbor Defense Board, comprised of the chiefs of coast artillery, engineers, ordnance, and air corps, and chaired by the chief of the Chemical Warfare Service, Major General Walter C. Baker, had resurveyed seacoast defense requirements. On 27 July, 1940, General Baker submitted the board's report to the chief of the General Staff.[27]

This report reiterated the views of previous boards that the 16-inch gun be the primary seacoast defense weapon and that the 6-inch gun be the secondary weapon in all fixed harbor defenses. It then proposed twenty-seven new batteries of casemated 16-inch guns and either casemating six 16-inch batteries built in the 1920s and 1930s or provided them with gun shields for overhead protection. The board further recommended another thirteen continental batteries of casemated long range 12-inch guns. The board went on to recommend fifty new batteries of 6-inch guns on long-range barbette carriages. Once these modern batteries were in service it would be possible, in board's view, to abandon 128 obsolete and obsolescent seacoast batteries. These new batteries could also replace the use of WWI-era railway artillery, along with widespread use of the WWI-era mobile 155-mm guns to fill the gaps until the new 6-inch batteries could be constructed.[28] The General Staff approved the modernization program on September 5, 1940, and recommended that 75 percent of the $82,666,000 modernization project be allocated for construction to be undertaken through the end of the 1942 fiscal year.[29]

The emplacements for these new batteries would differ from previous designs as protective measures from both naval shelling and aerial bombing were provided by adding reinforced concrete casemates and/or armored shields. The 16-inch guns were mounted in pairs inside thick concrete casemates approximately 500 feet apart. Between the casemates was a service tunnel off which extended a series of galleries which contained magazines, shell rooms, generators, and various storage and operating facilities. The entire structure was roofed by 8 to 10 feet of reinforced concrete, which in turn was covered with a layer of sand up to 20 feet thick. The guns in the casemates were further protected by armored shields and additional overhead concrete and steel. This standard design became known as the #100 Series Battery Construction. All the 16-inch guns used in the Delaware Bay defenses were US Navy guns which became available to the US Army as a result of the Washington Naval Conference of 1922 limitations on new warship building. This 16-inch gun had a range of 45,155 yards or about 25 miles. The new 6-inch guns had a thick wrap-around armored shield and they were mounted on open concrete

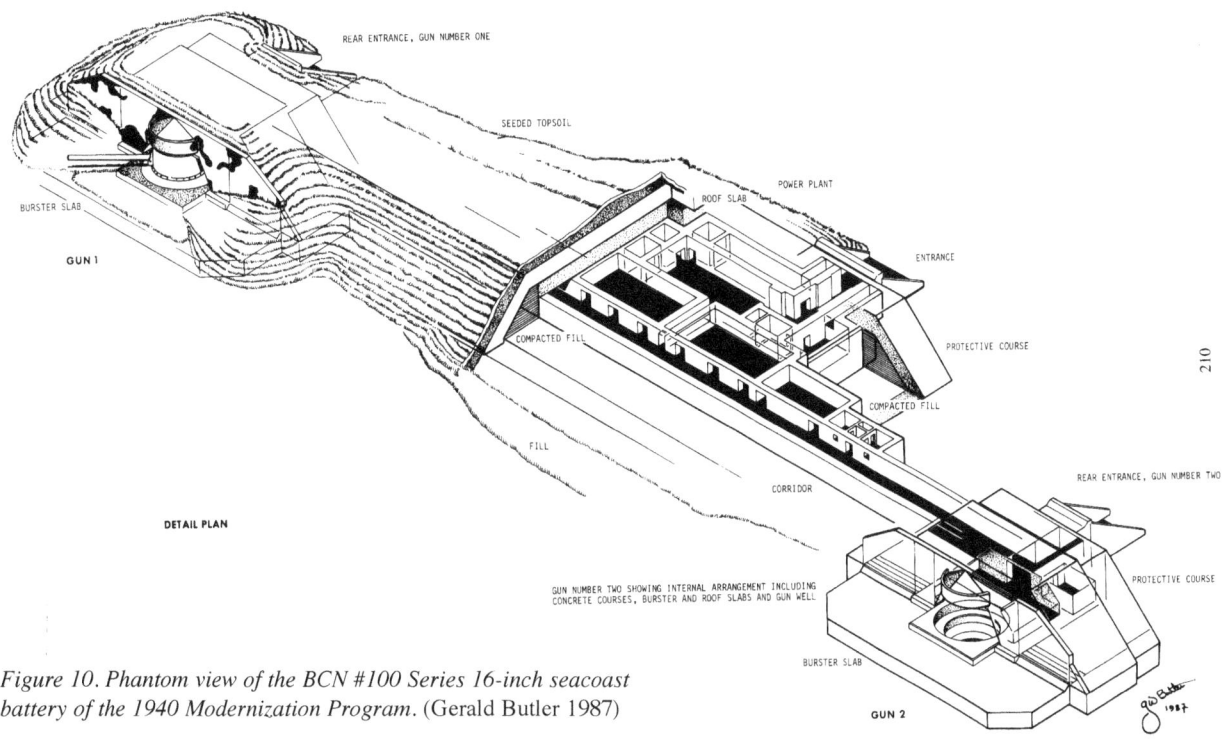

Figure 10. Phantom view of the BCN #100 Series 16-inch seacoast battery of the 1940 Modernization Program. (Gerald Butler 1987)

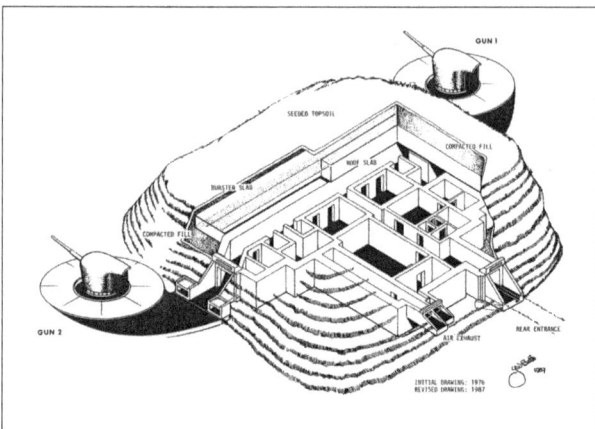

Figure 11. Phantom view of the BCN #200 Series shielded 6-inch battery of the 1940 Modernization Program. (Gerald Butler 1987)

pads, one on each side of the central traverse. Between the gun positions was an earth covered reinforced concrete structure containing the magazines, shell rooms, generators, and storage and operating space. This standard design became known at the #200 Series Battery Construction. These guns had a range of 27,530 yards or about 15 miles.[30]

The Harbor Defenses of the Delaware figured prominently in the Harbor Defense Board's modernization program. Batteries Hall and Haslet at Fort Saulsbury were to be casemated. Farther down the bay at Cape Henlopen, two batteries of 16-inch guns were to be built. One of these was the battery authorized in 1934. As construction of this battery had not yet begun, it was incorporated into the modernization program and was designated Battery Construction Number (BCN) 118. The second of the new batteries, BCN 119, was also to be built on the cape. In addition, the modernization program called for three batteries of long-range 6-inch guns on the capes. BCN 221 and 222 would be located at Cape Henlopen; BCN 223 was to be built at Cape May. Once the new batteries were in place, the older armament at Fort DuPont, Delaware, and Mott could, in the view of the board, be either relocated or scrapped.

On 10 October, 1940, the Adjutant General named a group of officers to a site board to work with the local boards in the Second Corps area. This board was composed of Col. Avery J. Cooper, CAC, president of the board, and Majs. Harry B. Vaughn, CE, and Paul A. Harris, CAC. About a week later, Second Corps named the officers of the local board that was to work with the site board to select locations for the modern batteries in the Harbor Defenses of the Delaware. This board was Colonel Earl Biscoe, CAC; Colonel Lucian B. Moody, Ord. Dept.; Colonel Frank B. Lahm, AC; Colonel John C. Moore, SC; Lieutenant Colonel James C. Hutson, CAC; Lieutenant Colonel Leight F. J. Zerbee, CE; and Major Harry B. Vaughn, CE.[31]

On 21 October, 1940, authorization for the removal of Battery Torbert's 12-inch guns at Fort Delaware was received at HQ, HD of the Delaware. Little time was lost in dismounting the massive guns from their disappearing carriages. By November, the guns were on their way to Watervliet Arsenal per G.O. No. 1268 of the War Department. After modification, the three guns were sent to Puerto Rico where two of the guns were mounted on long-range M1917 barbette carriages in Battery Reed at San Juan. The third gun tube was stored at the battery as a spare. Although the guns were removed from Battery Torbert, the three disappearing gun carriages remained in place until March 1943, when they were turned over to the harbor defense salvage officer and scrapped.[32] The removal of Battery Torbert's guns left only the 3-inch RF guns of Battery Hentig on Pea Patch Island. On February 6, 1941, transfer of Battery Harker's three 10-inch guns was authorized, and by March of that year, the guns had been removed from their emplacements. Two of the 10-inch M1888M1 guns were emplaced at Cape Spear, in the defenses of St. John's, Newfoundland. The third was emplaced at Fort Prevel on the Gaspe Peninsula of Quebec.[33]

Coastal Defenses Move to the Delaware Capes

By January 1941, the harbor defense project for the Delaware had been updated. This supplement called for the removal of the 12-inch guns of Battery Arnold at Fort Mott, while the 3-inch guns of Battery Hentig at Fort Delaware and Battery Elder at Fort DuPont were to be retained and provided with wraparound shields. Farther down the bay at Fort Saulsbury, plans were underway for casemating the guns of Batteries Hall and Haslet. Pending completion of the new defenses, a battery of four 8-inch railway guns and two four-gun batteries of mobile 155 mm guns were to be temporarily emplaced on Cape Henlopen. Although slated for deletion from the harbor defense project, Battery Arnold's guns remained in place, but unmanned, until its abandonment was authorized by the New York-Philadelphia Sector on August 31, 1943. The armament was not removed until sometime in 1944.[34]

In 1941, the Harbor Defenses of the Delaware were directed to establish a harbor entrance control post (HECP) at Fort DuPont, and the local joint planning board was charged with locating the installation. The command post at Fort DuPont was considered unsatisfactory, being too far up the bay. Initially, the board recommended that the HECP be at the navy section base at the Naval Air Station, Cape May. This location was so far from the main ship channel, however, as to be virtually useless, even as a temporary expedient. The joint planning board then contemplated a combined Harbor Defense Command Post (HDCP) and HECP in a single structure near the Great Dune on Cape Henlopen. By May 8, 1942, the board concluded that an HECP at the Great Dune would also be too far from the channels, making visual communication too difficult. Finally, on March 6, 1941, the board recommended a permanent joint army-navy command post adjacent to the army's gun battery and minefield control station at Cape Henlopen, when these elements of the defenses were constructed.[35]

Figure 13. Battery C, 261st C.A. Bn at firing practice with 155mm M1918M1 GPF gun on M1917 carriage in May 1941 at Cape Henlopen Military Reservation. (DNREC 1941)

Cape Henlopen was wind swept dunes spotted with sparse vegetation in the form of grasses and thickets of low beach heather. A grove of poplar trees stood at the site of the quarantine station, and inland from the high dunes were stands of willow and pine. Amidst some of the sections of dunes there were low lying tidal marshes covered in marsh grasses. On the beach about a mile and a half south of the tip of the cape was the Bell Haven Surf Club. This structure, initially built as a U.S. Life Saving Service station shortly after the turn of the century, had later served the U.S. Coast Guard. It had occupied a spot on the bay side of the cape for some 35 years until it was sold and moved to the oceanfront as a private beach club in 1939. It was one of the few buildings on the cape when the engineer survey teams arrived in 1940.

The Cape Henlopen Military Reservation had been established in 1938, initially encompassing the former site of the quarantine station (1884-1926) and the World War One 6-inch gun emplacement. Activity began to pick up at the reservation in the summer of 1940 when the Corps of Engineers surveyed. Soon afterward, the development of the site for modern harbor defenses got underway. The reservation was enlarged in 1940 until it encompassed over a thousand acres, extending some four miles southward from the point of the cape to near Gordons Pond and about a mile inland from both the ocean and the bay shore. (By the time Fort Miles was inactivated, it encompassed some 1,440 acres).[36]

On 16 April 1941 Maj. Ralph S. Baker, with a detachment of HQ and HQ Battery and Battery C, 261st C.A. Bn., deployed from Fort DuPont to Cape Henlopen. The majority of the detachment's personnel traveled by

Figure 12. Belle Haven Surf Club transformed into the temporary Harbor Entrance Control Post (HECP) for the Harbor Defenses of the Delaware at Fort Miles, DE taken on January 23, 1942. (RG77 NARA 1942)

truck in the light column and reached Cape Henlopen the same day. The heavy equipment of the detachment, four 155 mm M1918M1 GPF guns on M1917 carriages, two portable 60-inch searchlights, one M1 Caterpillar tractor, and the battery's fire control equipment, moved on the Pennsylvania Railroad and did not arrive in Lewes, Del., until 20 April. The coast artillerymen spent the next few weeks establishing their encampment on the cape and preparing firing positions for the guns. Initially the troops bivouacked under canvas in the rear of the guns. By 10 May the armament was in place and Battery C conducted a firing practice that same day between 0800 and 1100 to settle in the guns.[37]

In the latter part of 1940 and the early months of 1941, preliminary work on the modern batteries got underway. The initial activities involved letting contracts, collecting building materials, and preparing the sites for the first of the two projected 16-inch gun batteries. Although authorized in 1934, funding for BCN 118 was not granted until development of the Cape Henlopen Military Reservation began and the modernization program was developed. Work was initiated under contract by the White Construction Company and the firm of George and Lynch that had received the contract for construction of the fortifications as well as the numerous other structures on the reservation. The site for the 16-inch casemated battery was amidst the dunes several hundred yards inland from the ocean, about three miles south of Cape Henlopen Point on the southern fringe of the Great Dune. Work at the site began on 24 March 1941. The design of the battery was similar to others along the Atlantic seaboard that had been approved in 1934. The site was excavated down to only 6-inches above sea level. Then pilings were driven some sixty feet down to support the concrete

Figure 14. A 155mm M1918M1 GPF gun on M1917 carriage with its gun crew conducting firing practice at Fort Miles, DE in November 1941 with other 155mm GBF guns in background. (DNREC 1941)

foundations of the battery. Timber piles were used under the gun blocks, while piles of a composite material were used under the remainder of the battery structure. Construction continued for some twenty months before the battery was completed.[38]

The centers of the 43-foot square poured concrete gun platforms were 500 feet apart. Between them a massive reinforced concrete structure housing various service functions and ammunition storage rooms had been built. The casemates were connected by a magazine service corridor that extended the length of the traverse. Along the back wall of the main corridor were two storerooms, two shell rooms, and two powder rooms. At the mid-point of the corridor another corridor provided access to the power room, two more storerooms, an air-compressor room, water cooling plant room, latrine, and muffler gallery. At each end of the main corridor at the rear of gun casemates was a toolroom. There was a 16-foot wide entrance corridor at the rear of each of the casemates. Above the heavy concrete roof of the magazine/service traverse was a sand cushion covered by a two-foot thick concrete busters' course and about two more feet of earth and sand planted with vegetation so as to blend with the surrounding sand dunes. From floor level to the top of the "sand dune" cover, the battery was some 44 feet high.

The guns of Battery 118 were Navy MkIIM1 16-inch guns that had become available as a result of the warship cutbacks associated with the naval treaties of the 1920s. The guns allotted for the battery authorized in 1934 for Cape Henlopen had been stored at Aberdeen Proving Ground until the battery was ready for them. They were mounted on Army M4 barbette carriages.[39]

Concurrent with the construction of Battery 118, a separate concrete plotting and switch-board room for the battery was begun about 1,000 feet to the southwest of the battery. This gas-proof command center for the battery was also built on pilings and constructed of reinforced concrete. It was dug into the sand dunes and

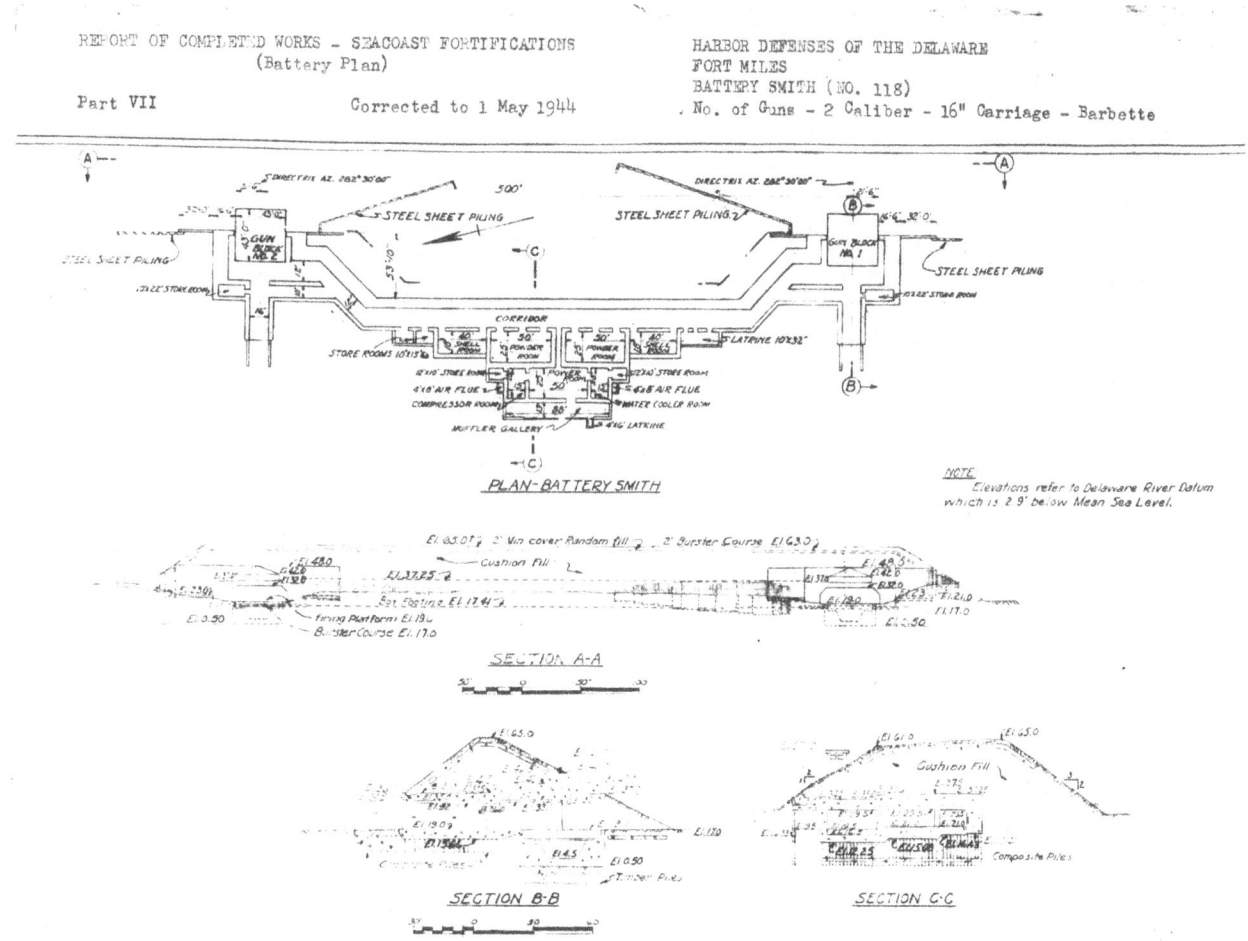

Figure 15. RCW Form 7 for Battery #118 showing the plan and sections for the 16-inch casemated battery (later Battery Smith at Fort Miles). (RG77 NARA 1944)

provided with a thick earthen cover over its 8-foot thick reinforced concrete roof, rendering it bombproof. The structure had two entrances some 73 feet apart on the building's landward side. The seaward walls of the PSR were also 8 feet thick, while those on the land faces were three to four feet thick. The three principal rooms in the

Figure 16. Construction underway at for Casemate #2 of Battery #118 at Fort Miles, DE on September 23, 1941 showing one of the large steel girders that reinforces the roof of the casemate being lifted into place while rebar and wooden forms are being assembled and a cement truck is making a delivery. (RG77 NARA 1941)

Figure 17. Casemate #1 of Battery #118 at Fort Miles, DE on January 24, 1942 showing the gun casemate with its protective canopy is nearly complete while in the background one can see a fire control tower and a reserve magazine in the distance. (RG77 NARA 1942)

building were the switchboard room, about 18 by 24 feet; the plotting room, 30 feet by 24 feet; and the spotting room, 10 feet by 24 feet. In the rear portion of the structure were rooms for the Signal Corps equipment and storage batteries, an air lock and a washroom associated with the gas-proofing, storerooms, an equipment room for the Chemical Warfare Service gas-proofing materiel, and latrines.[40] Although completed, BCN 118 was not

Figure 18. Rear view of Casemate #2 of Battery #118 at Fort Miles, DE on January 23, 1942 showing the large size of the gun casemate in relation to the truck parked by the rear entrance which itself is large to allow the 16-inch barrel to be moved into the gun casemate. (RG77 NARA 1942)

Figure 19. 16-inch/50 Mk2 barrel #44 on railcar at Cape Henlopen prior to transfer via truck to Battery #118 at Fort Miles, DE on July 9, 1942. (RG77 NARA 1942)

proof fired until about the middle of September 1943.[41]

In addition to the 16-inch battery, work was begun on mine storehouses, loading rooms, a wharf for the mine planting vessels, and other structures supporting the submarine mining operations planned for the bay. These facilities were located on the bay shore of the anchorage in the lee of the cape that was known as the harbor of refuge.

In June 1941, the decision was made by the local board to mount the four 155mm guns at Cape Henlopen on Panama mounts to be constructed just north of the Bell Haven Surf Club, and on 5 June Battery B, 261st C.A., deployed from Fort DuPont to the cape with four more 155 mm guns to be emplaced in field positions near those of Battery C. On 30 June the battery moved to Cape May Point, and an M1 Caterpillar tractor and 200 rounds of ammunition were sent down from Fort DuPont.[42]

While the initial steps to establish defenses at the entrance to the bay were underway during the spring and early summer of 1941, measures were also taken to reactivate the 12-inch long range gun batteries at Fort Saulsbury. In January 1941, Batteries Haslet and Hall had been slated for casemating.[43] Soon afterward, newly activated Battery C, 21st C.A., was sent to Fort Saulsbury to reactivate the guns. By 15 May, however, the Fort Saulsbury project was modified. All work on Battery Haslet was to be halted, its guns dismounted and relocated to Fort Miles.[44] Work on Battery Hall was to be continued, however. While initially Battery C lived under canvas at Fort Saulsbury, temporary theater of operations housing was constructed in time for winter.

At the end of June 1941, the projected tactical organization of the H.D. of the Delaware was to be the batteries noted in Table 2. While the three projected 6-inch barbette batteries were included in this proposed tactical organization, BCN 118 (2 x casemated 16-inch LRBC guns) was for some unknown reason not included in the tactical setup.[45]

Although some progress had been made, providing a semblance of defense at the Delaware Capes, the mine defense was almost non-existent in June 1941. There were no troops assigned to the mine command, and neither of the two mine planters required was available. There was no TNT for the mine charges; no mine yawls of the dozen that would be finally assigned; only one distribution box

Figure 20. Battery #118 Plotting and Switchboard Room (PSR) at Fort Miles, DE prior to the placement of protective earth and buster course as workmen start to apply waterproofing to the structure on September 23, 1941. (RG77 NARA 1941)

Figure 21. Wooden protective barrier is being built around the 12-inch gun #1 at Battery Hall, Fort Saulsbury, DE on April 29, 1941 in preparation for the construction of a casemate to protect the gun from naval shelling and aerial bombardment (this casemating project was soon cancelled). (RG77 NARA 1941)

boat, the L-56, of the four authorized; and no mine casemate. In August 1941, Batteries A and B, 21st C.A., travelled to Fort Hancock for submarine mine instruction, culminating in a mine practice.[46]

In addition to the tactical elements of the harbor defenses noted above, the 62nd C.A.(AA) was still training at Cape Henlopen. Battery A, 122nd C.A.(AA) Bn., was now stationed at Fort DuPont with four mobile 3-inch AA guns and fifteen serviceable 60-inch AA search-lights. Also at Fort DuPont were two 155 mm GPF guns without crews. Two 60-inch searchlights were serviceable at the entrance to the Delaware River. One, believed to be a portable light, was located at Fort DuPont to provide illumination for Battery Elder, the second, for Battery Hentig, was a fixed light on the southwest bank of Pea Patch Island atop a 65-foot disappearing tower. Both searchlights were manned by detachments from the 21st C.A.[47]

On 7 August, 1941, the War Department named the military reservation at Cape Henlopen as Fort Miles, in honor of General Nelson Appleton Miles.[48] While some news of the activity on Cape Henlopen had leaked to the newspapers during 1941, the naming of the new coast artillery post was about the first public mention by the War Department. Beyond noting that the original Sussex County reservation had been ceded to the United States in 1873, the army continued to withhold all further information concerning the fortifications or the organizations there.[49]

Progress on the 16-inch gun battery and the submarine mine facilities was temporarily halted on 26 August 1941, by labor disputes, and the union leadership called for a strike. The six hundred workers employed on the Cape Henlopen project set up pickets protesting the hiring of non-union painters and demanded a raise from $1.25 to $1.50 an hour. Although there was no disorder among the striking workers, the project manager for the contractors stated that "anybody who strikes will lose his job." Labor troubles would continue to plague the construction program at Fort Miles. Although wages for even common labor were higher than the norm, getting them to work was a frequent problem even as late as the middle of 1943. Rather than allowing the various gangs of common laborers to run the jobs their way, Major Lester L. Lessig, chief of the control branch of the Philadelphia Engineer District recommended that these men be terminated even if that action prolonged completion of the various projects.[50]

On 8 July 8, the local joint army-navy planning board decided that the temporary location of the HECP should be near the "Army battery [or] mine control station at

Table 2 Delaware Harbour Defences, June 1941

Battery	ID	Location	Armament	Status
Tactical Battery No. 1	BCN 221	Cape Henlopen	2 x 6-inch guns on long range barbette carriages	Under construction
Tactical Battery No. 2	BCN 519	Cape Henlopen	2 x casemated 12-inch guns	Under construction
Tactical Battery No. 3		Cape Henlopen	4 x 155 mm GPF guns	Emplaced. Battery C, 261st C.A.
Tactical Battery No. 4	BCN 222	Cape Henlopen	2 x 6-inch guns on long range barbette carriages	Under construction
Temporary Battery No. 4		Cape Henlopen	4 x 155 mm GPF guns	Battery B, 261st C.A., to be moved to Cape May
Tactical Battery No. 5	BCN 223	Cape May	2 x 6-inch guns on long range barbette carriages	
Tactical Battery No. 6		Battery Hall, Fort Saulsbury	2 x 12-inch guns, no fire control stations	Being placed in service by Battery C, 21st C.A.
Tactical Battery No. 7		Battery Elder, Fort DuPont	2 x 3-inch guns	Battery A, 21st C.A., manned around the clock
Tactical Battery No. 8		Battery Hentig, Fort Delaware	2 x 3-inch guns	Battery A, 21st C.A., manned around the clock
Tactical Battery No. 9		Battery Arnold, Fort Mott	2 x 12-inch disappearing guns	Unmanned, but otherwise ready for service, pending inactivation.

Cape Henlopen because of the poor visibility [of the channel] from Cape May."[51] On the same day that the construction workers went on strike, the first temporary signal station and HECP was established atop a wooden tower adjacent to one of the temporary wooden buildings newly constructed on Fort Miles. This structure was located just inside the bay overlooking the harbor of refuge. The army personnel assigned to the temporary HECP were housed in tents at the site. The wooden signal tower would serve until the better located Bell Haven Surf Club building could be adapted for use as the temporary HECP. Regulations for entering and departing the bay were promulgated by the Harbor Defenses of the Delaware and the station was placed in operation on a training basis 26 August.[52] On 16 October, after the temporary station had been operating in a training mode some six weeks, the commanding officer of the Fourth Naval District condemned the "make shift set up" of the temporary HECP and signal station as "unsatisfactory." In spite of his poor opinion of the harbor command post, the HECP went on continuous alert on 28 October, 1941. The navy personnel assigned to the HECP were commanded by Lt. Cmdr. Frank S. Carter, USN.[53] The former Naval Radio Compass Station, Cape Henlopen was reopened for direction finding (DF) duties and to provide administrative support facility for the HECP in the former Surf Club building.

In November 1941, the US Coast Guard was placed under the operational control of the US Navy. This meant the Commander of the Lewes Coast Guard Station, Capt. E.A. Coffin, became the local Captain of the Port (COTP).[54] The COTP was responsible for controlling the movement and anchorage all ships and the wartime function of the safety and security of ports and harbors and the formation of convoys. The HECP detachment would contact incoming ships by the time they were 12 miles offshore, identifying them, directing them to an examination anchorage, and insuring they stopped for examination. If a ship was not identified or did not stop, the HECP was responsible for informing an Army "examination battery" ashore to fire an un-fused "plugged round" to encourage the ship to stop. Failure to stop would result in the activation of the harbor defense mines and batteries. The COTP would send a station ship to the examination anchorage and board an examining officer and party to determine the "friendly character and intentions" of the ship prior to clearing it to enter the port. At the same time, pilots would board and ship would be given port orders and rules, and an armed guard could be put on board if required. The HECP would be informed of clearance and they would inform the harbor defenses to stand down.[55]

The initial construction program at Fort Miles had focused on structures directly associated with the fortifications. The planned cantonment for the troops assigned to the defenses received a lower priority. As the summer of 1941 ended, the National Guardsmen's hopes that they would return to their civilian pursuits at the end of the "year of training" had been dashed, and federalization of the National Guard had been extended indefinitely. Faced with the prospect of a winter on Cape Henlopen, the men of the 261st C.A. found their tent city less and less to their liking. Although on 27 September 1941, they were advised that a $500,000 construction program, including ten barracks and four mess halls of the mobilization type, would begin "very soon," the men of the 261st remained unhappy with progress on the quarters then under construction at Ft. Miles.[56] At first, progress had been slowed by material shortages and later by labor disputes. To ease the situation on the cape, the harbor defense command procured special stoves to be installed in the tents if necessary, in an effort to make the tent camp safe and comfortable.[57] This news fell on deaf ears, however, and by the middle of October pressures increased to rotate the guardsmen at Fort Miles back to Fort DuPont, and replace them with the Regular Army 21st C.A.[58]

With labor disputes resolved, at least temporally, advancement of the defenses continued. The casemates of BCN 118 were steadily progressing, and on October 14, 1941, it was named in honor of Maj. Gen. William R. Smith, USA, who had died in July.[59] The $500,000 barracks construction program at Fort Miles began in November. Ten mobilization type barracks, for 440 men,

Figure 22. The soldiers of 261st C.A. Bn (a former DE National Guard unit) lived in these winterized tents at Fort Miles during 1941 and 1942 at Fort Miles, DE until barracks could be built. (DNREC 1941)

and four mess halls were finally begun. Upon completion of this garrison program there would be theater of operations barracks for an additional 1,926 enlisted men, bachelor officers' quarters for fifteen officers, and quarters for ninety more officers in theater of operations structures, along with a 90-bed hospital (located in the town of Lewes) and an infirmary, in the cantonment area of the Fort Miles.[60]

As relations between the United States and Japan continued to worsen in late November and early December 1941, Colonel Ruhlen ordered Fort Miles closed to casual civilian visitors on 1 December. Although use of the Bell Haven Surf Club by bathers had ceased several months before, the closing of the reservation officially closed the club.[61] During the late autumn of 1941, the building had undergone extensive renovations and modification, for which the sum of $10,900 had been allocated, and on 4 December 1941, the Bell Haven Surf Club was occupied by the Army and Navy as the new temporary HECP, from which they continued around the clock oversight of the ship traffic in and out of Delaware Bay.[62]

War Comes to the Delaware Bay

At 5:45 p.m. on 7 December 1941, about four hours after the Japanese air raid on Pearl Harbor, the 2nd Coast Artillery District sent a terse three-word message to Colonel Ruhlen at Fort Miles: "Condition II Immediately."[63] This brief alert called for all harbor defense and antiaircraft artillery observation stations and communications to be manned around the clock. Further, at least one major caliber battery and one secondary battery, along with the necessary searchlights at night, were to be manned and ready to fire. All other batteries were to be fully manned and ready to fire within three minutes during daylight and five minutes at night.[64] The Harbor Defenses of the Delaware were now on full alert.

With the nation at war, the security of Delaware Bay became a matter of the utmost importance. Strategically, much of the oil consumed in the northeast quarter of the country was refined or otherwise processed in the Delaware River Valley between Wilmington and Philadelphia. The Philadelphia Naval Shipyard, with numerous warships currently under construction, was one of the navy's most important installations. Additionally, the chemical and munitions plants along the length of the Delaware Valley constituted a principal source of

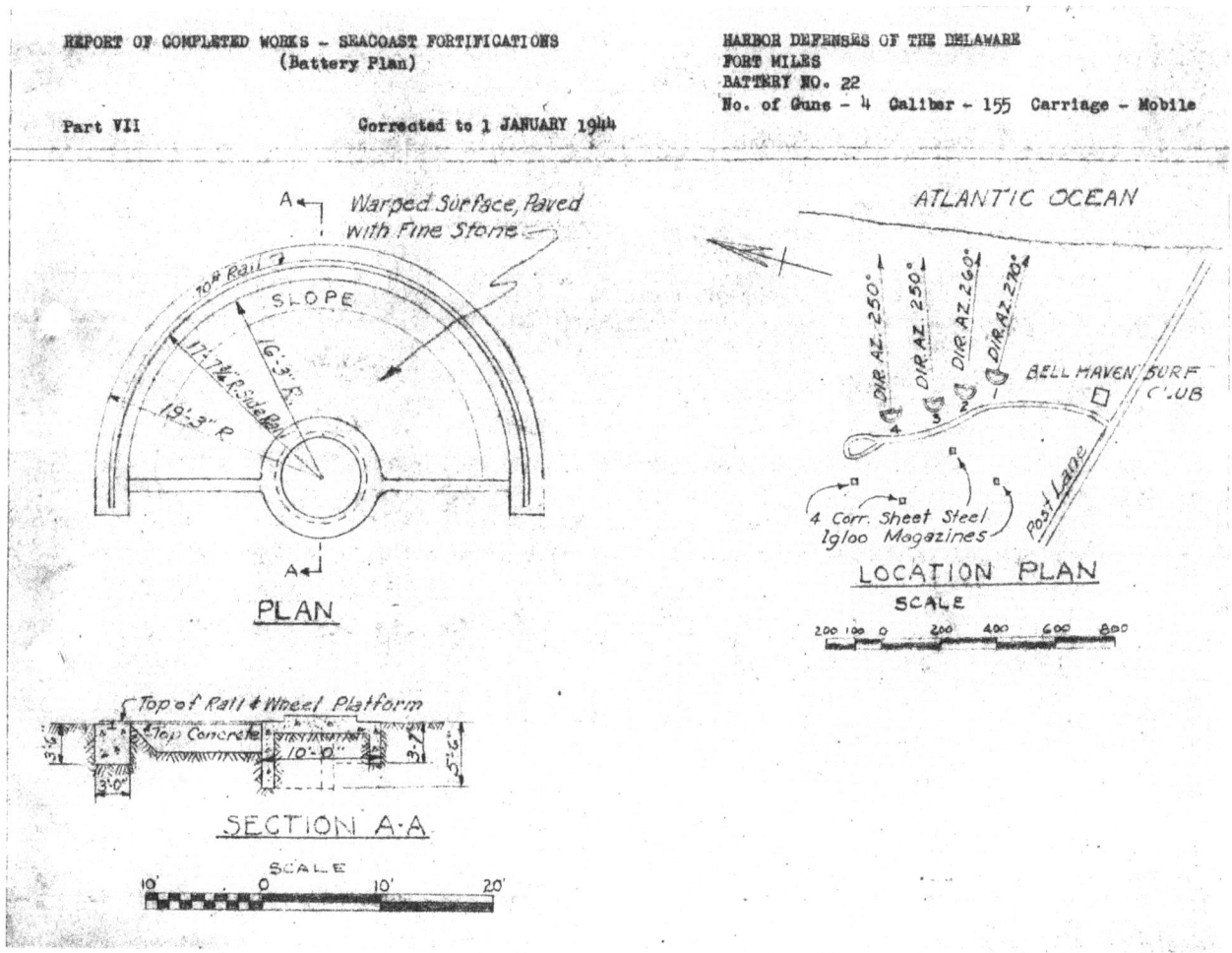

Figure 23. RCW Form 7 for Battery #22 showing the plan, section, and location plan for the 155mm GPF Panama mounts at Fort Miles, DE. (RG77 NARA 1944)

munitions for the army and navy, not to mention America's allies.

The entry of the United States into the world-wide conflict had an immediate effect on the development of Fort Miles. The pace of construction quickened on nearly all incomplete projects, both permanent and temporary. Within a week of the Japanese attack, construction began on the four Panama mounts, authorized months before, for the 155 mm GPF guns just north of the Bell Haven HECP. These mounts were rushed to completion by 5 January 1942. This battery, now tactically designated Battery No. 22 in a revised and updated harbor defense project, functioned in conjunction with the HECP as the examination battery for the harbor defenses. Battery 22 was also the only seacoast battery in service on Cape Henlopen at that time. On 17 December Battery C, 21st C.A., moved from Fort Saulsbury to Battery No. 25 on Cape May, the four 155 mm GPF guns now emplaced on the beach property of the University of Pennsylvania east of the Coast Guard light station at Cape May. The battery personnel took up quarters initially in nearby St. Mary's convent. Their four Panama mounts were begun on 31 December 31, 1941, and completed on January 9, 1942. Administratively, the designation of Forts Miles and Saulsbury as sub-posts of Fort DuPont was changed by G.O. No. 25, HQ, H.D. of the Delaware, which made them separate posts of the harbor defenses on 20 December 1941.[65]

The sounds of war were not long in coming to the Delaware Bay defenses. About 10:45a.m. on 22 December, "heavy detonations [were] heard off-shore at Fort Miles and other locations up and down the coast." Major Ralph S. Baker, at Cape Henlopen, stated that "firing was heard at scattered points along the shore clear down to Maryland. There were no visible indications of gunfire but the reports were very clear." More of these detonations were heard shortly after noon. While the Navy

Figure 24. Construction underway on January 23, 1942 for Panama Mount #2 for Battery #25 at the Cape May Military Reservation, NJ. (RG77 NARA 1942)

had no comments, it is believed that the explosions may have been depth charge attacks on suspected German U-boats, or, even more likely, gunnery exercises by elements of the U.S. Atlantic Fleet.[66] Whatever the source of the firing, it quickly brought home the reality of war to the coast artillerymen in their battle positions amidst the dunes of the Delaware Capes.

Another action that measured the importance attached to the Delaware was the deployment of mobile supports a scant week after war was declared. From his headquarters at Fort Hancock, Colonel H. Norman Schwarzkopf, commander of the 113th Infantry, a former N.J. National Guard regiment, directed the regiment's 2nd Battalion to the central Delaware area on 16 December. The battalion established its headquarters at the Civilian Conservation Camp at Georgetown, some 20 miles west southwest of Fort Miles. Company F of the battalion was initially posted to the navy section base at Cape May for a few days before leaving for a new station at Cold Spring, N.J., on 18 December. These support troops remained in the Georgetown area for the most part until early February 1942, when the five infantry companies of the 2nd Bn., 113th Inf. were redeployed as follows: [67]

– HQ and HQ Company, Georgetown, Delaware
– Company E, Bethany Beach, Delaware
– Company F, Tuckahoe, New Jersey
– Company G, Port Republic, New Jersey
– Company H, Dias Creek, New Jersey

By 23 December, Colonel Ruhlen had reassigned his forces to man the available armament. While he continued to maintain harbor defense headquarters at Fort DuPont for the time being, most of his coast artillery command was to be deployed to the Delaware Capes as follows:[68]

– HHB, HD Del., Ft. DuPont
– Battery A, 21st CA, Mine Command, Fort Miles
– Battery B, 21st CA, Mine Command, Fort Miles
– Battery C, 21st CA, Battery No. 25, four GPFs, Cape May

Figure 25. Construction of mine cable storage tanks on July 23, 1941 along with two loading rooms and mine wharf (located between them) at Fort Miles, DE. (RG77 NARA 1941)

Figure 26. Construction of mine storehouse on September 24, 1941 at Fort Miles, DE use steel frame and asbestos panels. (RG77 NARA 1941)

- Battery A, 261st CA, Battery No. 22, four GPFs, Fort Miles
- Battery B, 261st CA, Battery Hall, two 12-inch, Fort Saulsbury
- Battery C, 261st CA, Batteries Hentig, two 3-inch, Fort Delaware; and Elder, two 3-inch, Fort DuPont

Implementing the submarine mine defenses of Delaware Bay was one of the most important tasks for the harbor defenses. The importance of keeping German submarines from the bay or its approaches was not lost on Colonel Ruhlen and his staff. Battery A, 21st C.A., was relieved from the 3-inch rapid-fire guns of Batteries Elder and Hentig on 1 January 1942, and sent to Fort Delaware, where it and Battery B of the 21st initiated the mine defenses. Battery C, 261st C.A., moved from Cape May to Fort DuPont and took over the batteries at Forts DuPont and Delaware. As soon as these moves were completed, the mine command was formally established per G.O. No. 3, HQ HD of the Delaware. The initial plan was to implement the underwater defense section of the 1933 project, calling for 25 grand groups of submarine mines across the main channel into the bay.[69]

On 9 January 1942, four days after the mine command was established, the engineers reported that the mine facilities begun early in 1941 were ready. These facilities, just inside the Fort Miles reservation on the shore of Breakwater Harbor, consisted of the mine wharf, the boathouse where the small craft were housed, two mine loading rooms, the mine storehouse, two small magazines, four igloo-type TNT magazines, four cable huts, the cable tanks, and the mine casemate. In addition, U.S. Army Junior Mine Planter *Colonel Henry R. Casey* was placed in service. A river freighter on the Delaware prior to the war, she had been converted into an improvised mine planter.[70]

Initially, the minefield consisted of two lines of controlled mines across the main channel from Cape Henlopen to Hen and Chicken Shoal. These mines were older M1 types, sphero- cylindrical, some forty inches in diameter.[71] They were laid in grand groups of nineteen mines. Each of these two lines of mines had 20 or more grand groups. "Each mine was connected to a distribution box with a single cable from each such box to the shore casemate. The mines were placed at controlled depths and held in position by a cable to an anchor. From the mine casemate control board, the mines could be tested for condition of their firing devices, set to signal the board when disturbed, set to fire when hit by a ship, fired selectively from a plotting board when a ship was plotted in the minefield."[72]

The junior USAMP *Casey* was the only mine planter in

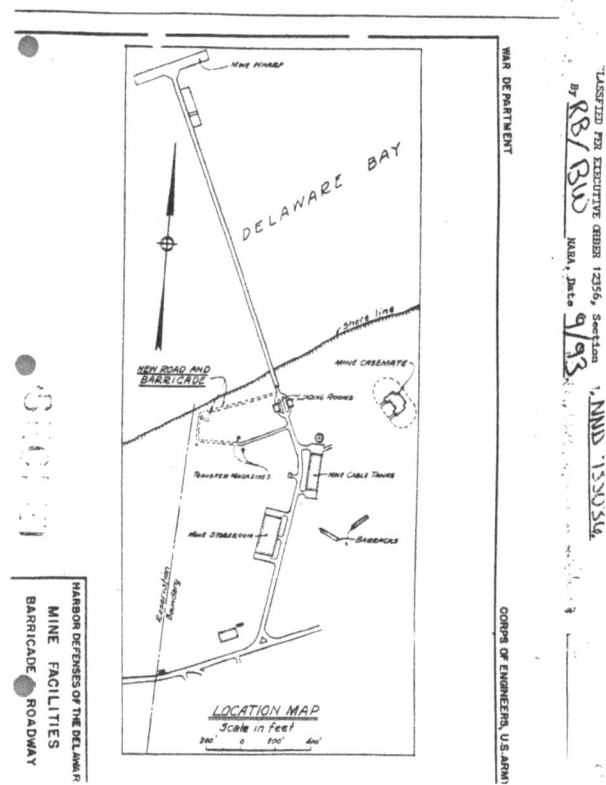

Figure 27. Location plan for the new controlled mine facilities (storehouse, cable tanks, loading rooms, magazines, casemate, barracks, wharf) at Fort Miles, DE. (RG177 NARA 1945)

Figure 28. US Army Mine Planter Schofield *prepares to drop a M1 mine and anchor as part of the Delaware Bay's mine defenses in April 1942. (NARA 1942)*

the harbor defenses until mid-March, when USAMP *General John M. Schofield* arrived from Fort Monroe.[73] The *Schofield* remained only until mid-April, when it returned to Fort Monroe.[74] The harbor defenses remained dependent upon mine planters from other harbors for most of 1942. After the departure of the Schofield in April, the junior mine planter *Casey* was responsible for maintaining the minefield until USAMP *General E.O.C. Ord* arrived

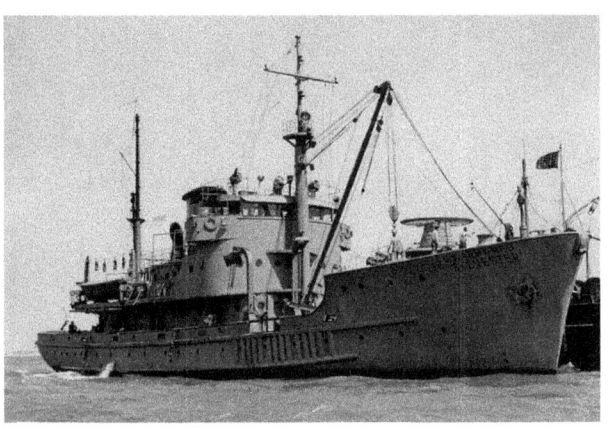

Figure 29. The newly commissioned USAMP Lieutenant William J. Sylvester *(MP-5) arrived at Fort Miles, DE on November 1, 1942 as the first permanently assigned mine planter to Harbor Defenses of the Delaware and to plant the new M-4 ground minefield.* (DNREC 1942)

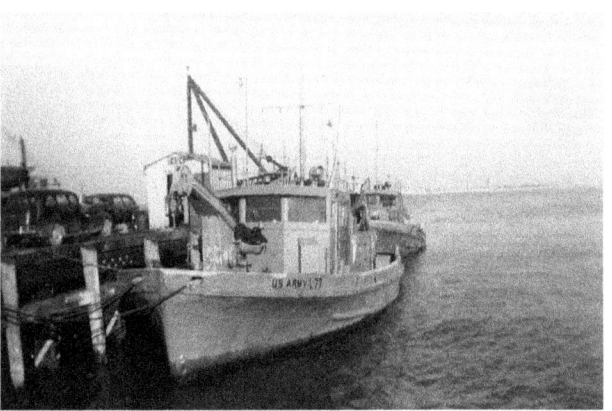

Figure 30. The mine distribution boat (L-77) *was one of four DB boats assigned to Harbor Defenses of the Delaware allowing for installation of the distribution hubs that connected the mine cables from each mine to the main cables to the mine casemate.* (DNREC 1944)

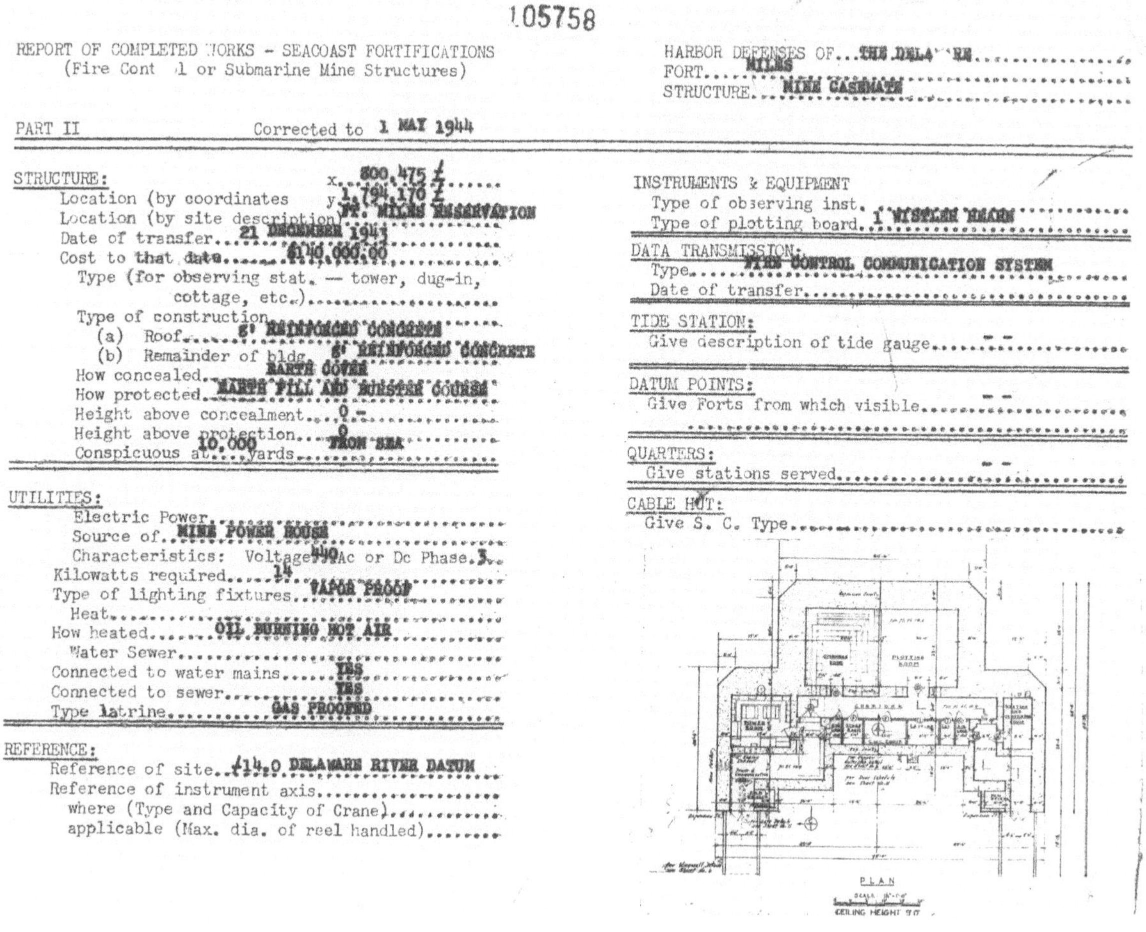

105758

REPORT OF COMPLETED WORKS - SEACOAST FORTIFICATIONS
(Fire Control or Submarine Mine Structures)

HARBOR DEFENSES OF... THE DELAWARE
FORT... MILES
STRUCTURE... MINE CASEMATE

PART II Corrected to 1 MAY 1944

STRUCTURE:
Location (by coordinates x... 800,475 ± y... 1,794,170 ±
Location (by site description) FT. MILES RESERVATION
Date of transfer... 21 DECEMBER 1941
Cost to that date... $140,000.00
Type (for observing stat. — tower, dug-in, cottage, etc.)...
Type of construction...
(a) Roof... 8' REINFORCED CONCRETE
(b) Remainder of bldg. 8' REINFORCED CONCRETE
How concealed... EARTH COVER
How protected... EARTH FILL AND BURSTER COURSE
Height above concealment... 0 -
Height above protection... 0
Conspicuous at 10,000 yards... FROM SEA

UTILITIES:
Electric Power...
Source of... MINE POWER HOUSE
Characteristics: Voltage 440 Ac or Dc Phase 3
Kilowatts required... 14
Type of lighting fixtures... VAPOR PROOF
Heat...
How heated... OIL BURNING HOT AIR
Water Sewer...
Connected to water mains... YES
Connected to sewer... YES
Type latrine... GAS PROOFED

REFERENCE:
Reference of site... +14.0 DELAWARE RIVER DATUM
Reference of instrument axis...
where (Type and Capacity of Crane)...
applicable (Max. dia. of reel handled)...

INSTRUMENTS & EQUIPMENT
Type of observing inst. 1 [illegible]
Type of plotting board...

DATA TRANSMISSION:
Type... FIRE CONTROL COMMUNICATION SYSTEM
Date of transfer...

TIDE STATION:
Give description of tide gauge... - -

DATUM POINTS:
Give Forts from which visible... - -

QUARTERS:
Give stations served... - -

CABLE HUT:
Give S. C. Type...

Figure 31. RCW Form 2 with plan for the Mine Casemate at Fort Miles, DE. (RG77 NARA 1944)

from New York on 10 August 1942. The *Ord*, launched in 1909, was of the same class as the *Schofield*, and remained until August 28th before returning to New York. The next day USAMP *General Henry J. Hunt* arrived from Narragansett Bay for about twenty days of mine planting, returning to Narragansett Bay on 20 September. The *Hunt*, MP-2, was second of the new class of sixteen 188-foot mine planters, the first of which had been laid down in 1941 at the Marietta Manufacturing Co., Point Pleasant, W.V. A month after the departure of the *General Henry J. Hunt*, the mine project for Delaware Bay was revised, calling for 35 groups of M4 ground mines in three rows in the deep water northeastward across the channel from Cape Henlopen to Hen and Chicken Shoals, some 8,000 yards. The M4 mine introduced in 1942 was a nine-foot cylinder some 90 inches in diameter, with a conical top. It contained some 3,000 pounds of TNT, and "was effective at depths of sixty to seventy feet against ships of any size or type."[75]

In addition to the new mines, three army hydrophones

Figure 32. Entrance to the Mine Casemate at Fort Miles, DE is uncovered (using a backhoe to remove the earth covering the door) in 2015 to allow the CDSG annual conference attendees to explore the casemate. (McGovern Collection 2015)

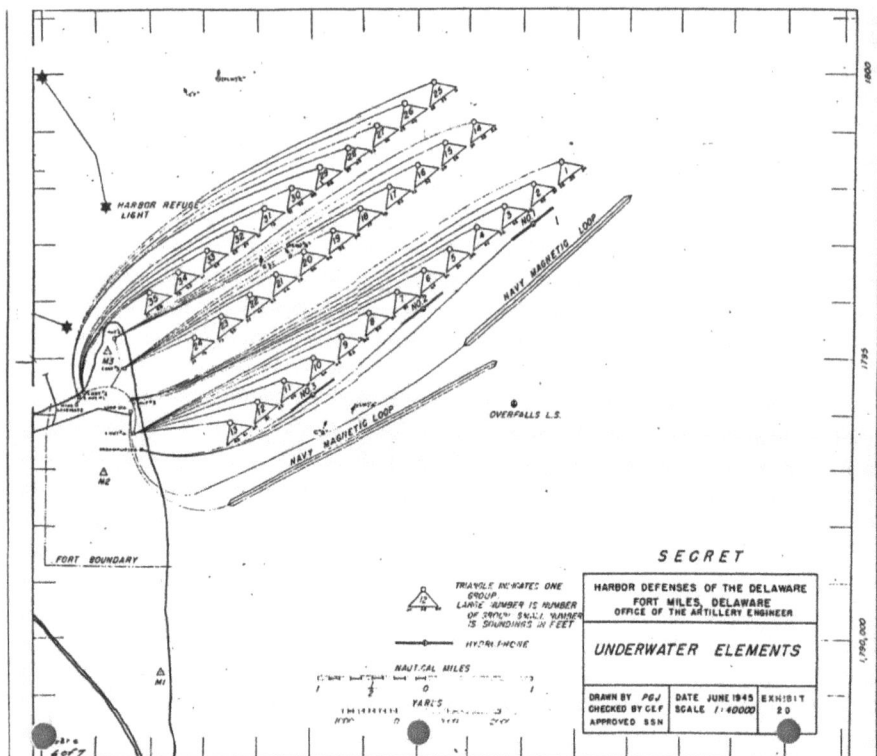

Figure 33. Location plan for final underwater elements of the Harbor Defenses of the Delaware which included 455 M-4 ground mines in June 1945. (RG177 NARA 1945)

were to be set in advance of the minefield. In advance of the hydrophones, two underwater magnetic detector loops operated by the Navy were to be established.[76] Near the end of the war, the minefield was again changed when M3A1 mine cases were provided that differed from the original M3 cases used with the M4 mines by having a 400-pound weight in the base to make them more stable and less susceptible to tidal action. These mines were re-laid in three lines, but the 35 grand groups of the mine field were composed of only 13 mines per group. The outer line was composed of 13 groups, while the middle and inner lines were comprised of 11 groups each.[77]

The first mine planter permanently assigned to Delaware Bay, the newly commissioned USAMP *Lieutenant William J. Sylvester* (MP-5), manned by the 12th C.A. Mine Planter Battery, arrived on 1 November 1942. The *Sylvester*, named for the first coast artillery officer to be killed in action in World War II, was the fifth mine planter from Point Pleasant. The 12th C.A.M.P. Battery was activated at Point Pleasant on 19 August 1942. *Sylvester* was joined by USAMP *Brigadier General Royal T. Frank* (MP-12) and the 19th C.A.M.P. Battery on 1 April 1943. The 19th C.A.M.P. Battery had been activated at Fort Hancock on 28 November 1942. Soon thereafter, it had been transferred to Point Pleasant to man the newly launched mine planter. The *Frank* was the second mine planter to bear that name. The earlier *Royal T. Frank* had served in the Hawaiian Department during World War I and following that war had been used as the army's inter-island transport until sunk by a Japanese submarine with considerable loss of life. Both the *Sylvester* and the *Frank*, along with the Junior Mine Planter *Casey*, served in the Harbor Defenses of the Delaware through the end of the war.[78]

In addition to the mine facilities transferred to the garrison on January 9, 1942, a searchlight shelter for four 60-inch M1941 Sperry portable searchlights, their Sperry controllers, and General Electric mobile power generator sets was also completed. The 110 feet long and 24 feet wide shelter was a steel frame structure faced with corrugated asbestos siding. The concrete floor and the asbestos roof made it nearly fire proof. At the same time, five concrete silo-type fire control towers were placed in service as base end stations at Fort Miles. On the west shore of Delaware Bay between Lewes and Fort Saulsbury, five steel-frame base end towers, one 48-foot and four 94-foot, provided target azimuth data to the batteries. Each of the towers had a three-tiered observation booth at its top, 11 feet wide and 10 feet 6 inches deep, and each tier had a ceiling of about 7 feet. At each fire control location, a reinforced concrete cable hut was also built. A searchlight shelter, 30 feet by 24 feet, of asbestos siding on a steel frame, was also completed and placed in service at Fort Saulsbury.[79]

Meanwhile at Fort Miles, the driving of pilings for BCN 221 began south of Battery Smith, about halfway to Gordon's Pond. Like Battery Smith, the foundations of the gun platforms were supported on pilings driven some 60 feet deep. Construction of this long range, 6-inch battery was begun on 15 January 1942. From this location nestled among the sand dunes, its weapons could cover the approaches to the bay for some fifteen miles to seaward. While the two guns of the battery were mounted in the open on long range barbette carriages 210 feet apart, they were provided with splinter proof steel wrap-around shields. The battery's magazines, storerooms, gas proofed plotting room, generator room, etc., were housed in a central traverse magazine of reinforced concrete between the two guns. This structure had a thick earthen cover topped with an eighteen-inch-thick burster course of concrete. An additional foot or two of earth and sand was then placed on top of the burster course, to about 60 feet above sea level. On top of, and partially dug into, the top surface of the structure was a small splinter proof battery commander's station of reinforced concrete. The battery was completed at the end of August 1943.[80]

Early in 1942, additional measures for the protection of shipping in the upper reaches of Delaware Bay were considered, and the Navy decided that an anti-motor torpedo boat boom and net should be established at Reedy Island. The local joint board recommended a gun battery and a 60-inch searchlight to protect the boom. On 10 March 10 1942 Battery Elder's pair of 3- inch rapid-fire guns and a 60-inch portable searchlight were moved from Fort DuPont to the Liston Front Range Light at Bayview, Del., where they were emplaced to cover the boom installed on 16 March. On 22 May G.O. No. 9, HQ Philadelphia Sub-Sector, relieved Battery C, 261st C.A. Bn., from Batteries Elder and Hentig, and moved the unit down to Battery No. 5's four 3-inch guns at Fort Miles. A detachment of Battery C manned the Reedy Island AMTB Battery, also known as Battery Liston and designated tactical Battery No. 7. The boom and net were maintained by the Navy only until 2 December 1942, when it was dismantled as it was decided that resources required to maintain this defense was greater than the benefit. On 12 December 1942, the armament of the battery was ordered removed, and on 10 March the harbor defense ordnance officer placed the armament in storage at Fort Saulsbury pending shipment to an arsenal.[81]

On 4 April the local joint planning committee sited a battery of four 3-inch guns to protect the anchorage between the inner and outer breakwaters and the moorings in Breakwater Harbor. This battery, Tactical Battery No. 5, also covered the minefields at the entrance to the bay. On 15 May construction began on the battery and continued until 31 August. In the meanwhile, two 3-inch rapid-fire guns and their shielded pedestal mounts were removed from Battery Turnbull, Fort Wadsworth, and shipped to Fort Miles on 23 May 1942. The other pair of guns, from Battery Hentig at Fort Delaware, moved to Fort Miles on 7 June 1942.[82] The four concrete gun blocks for Battery No. 5 were completed on 31 August 1942. The installation of the four M1903 rapid-fire guns on their shielded pedestal mounts was completed on 11 September 1942, and the battery was placed in service.[83]

On 18 February 1942 the HECP was assigned control of the fleet tug USS *Allegheny* (AT-19) as a station ship at the

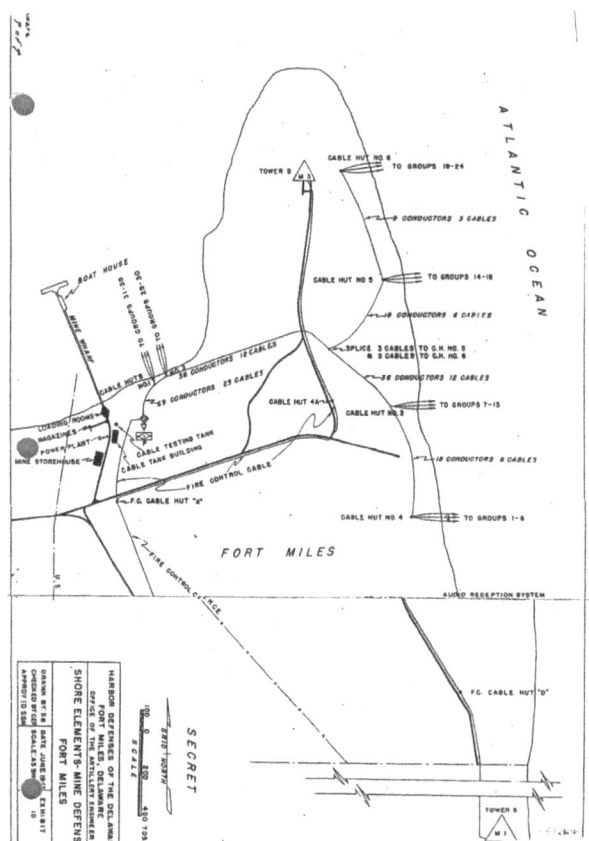

Figure 34. Location plan for the final shore elements of the mine defense for the Harbor Defenses of the Delaware in June 1945 with the numerous cable-runs and huts connecting with mines offshore. (RG177 NARA 1945)

entrance to the channel leading into bay with the 75-foot *Coast Guard Cutter 653* assigned as relief. That area was maintained as a mine-swept anchorage by a group of coastal minesweepers (AMc) known as the "little sisters" based at Cape May.[84] The converted yacht, USS *Alabaster* (PYc-21) with *YP-221* as a relief, was based at the Overfalls lightship to discourage submarine attacks on that ship, but also served to direct arriving ships toward the USS *Allegheny* and the anchorage. That sometime required the USS *Alabaster* to give a merchant ship a nudge or fire a shot across the offending ship's bow. Later, as the COTP expanded it role, the USS *Allegheny* was replaced by two Coast Guard tugs, the *Naugatuck* (WYTM-72) and *Yankton* (WYTM-92).[85]

In April 1942, the laying of anti-submarine magnetic loops at the entrance of the Bay completed by the contract ship Cyrus Field. The HECP underwater detection facility or loop station was built adjacent to the HECP as the termination point for the signals from these loops. There were two loops, each consisting of three cables strands about 5,000 yards long and spaced 200 yards apart, totaling 9.8 miles of cable. They were placed just inside the position of the Overfalls Lightship and just outside the outer row of the Army controlled minefield in deep water area extending between the Overfalls South Shoal and the Hen and Chicken Shoal. When a large metal object such as a ship or submarine passed over the loops it caused a magnetic imbalance between the cables which generate currents that were carried to the shore by the center stand. These were read by a galvo meter or fluxmeter monitored by an operator in the loop station. In addition to the magnetic loops, the HECP had a least three or four sono-buoy set for the detection of the sound of approaching submarines. It consisted of a buoy containing a radio transmitter and antenna, and from which a hydrophone was suspended down to a depth of about 60 feet. The buoy was powered by a battery contained in a separate battery raft to which it was connected by a wire rope and by an

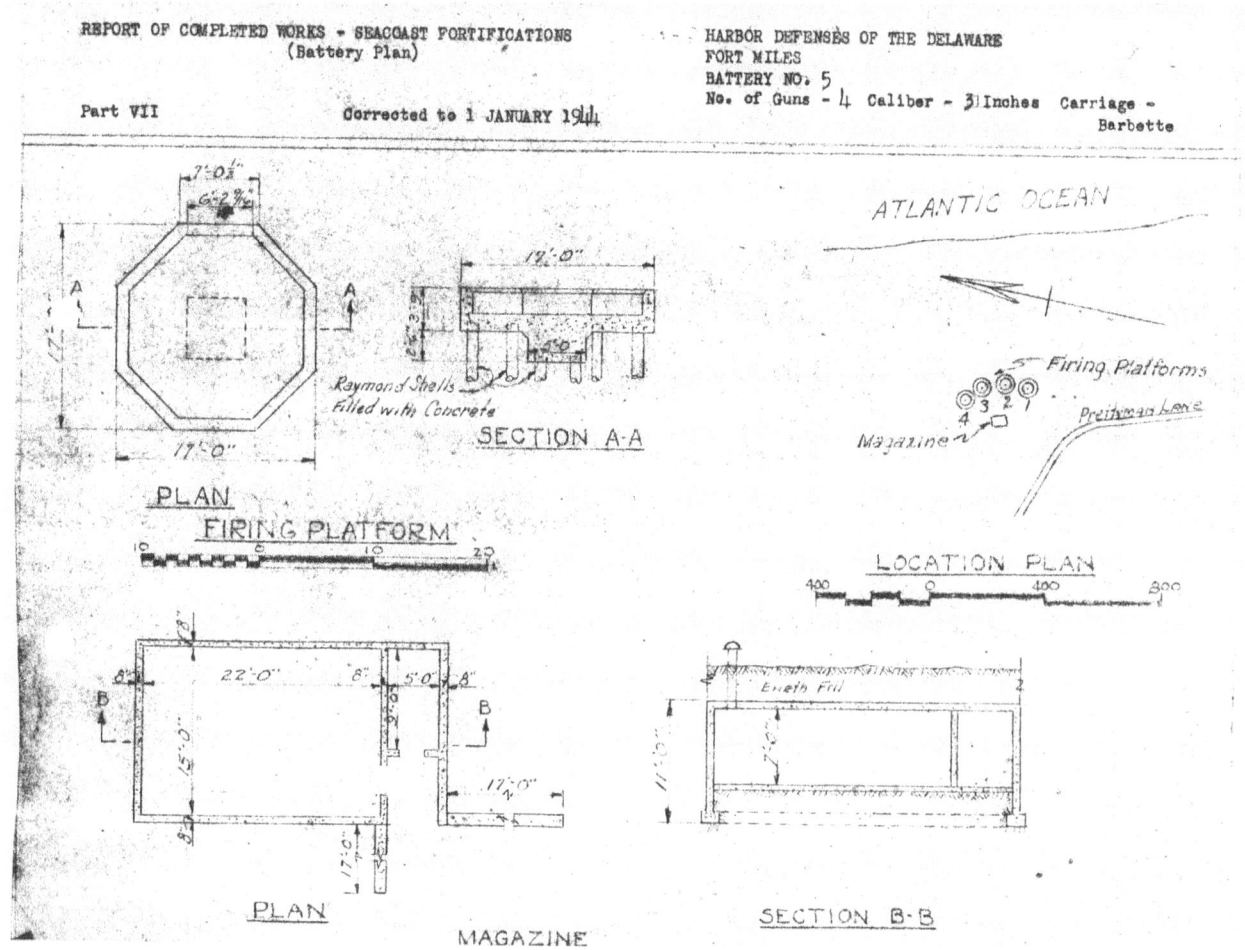

Figure 35. RCW Form 7 for Battery #5 showing the firing platform and magazine plan and location plan at Fort Miles, DE for the relocated 3-inch rapid fire guns to provide protection for the controlled minefield. (RG77 NARA 1944)

3-inch M1903 on M1903 Pedestal Mount at Battery Hentig, Fort Delaware, DE before the battery's two guns were moved to Battery #5 at Fort Miles, DE on June 7, 1942. (DNREC 1941)

electrical power cable. The whole arrangement was anchored to the bottom by a cable from the battery raft. The hydrophones could detect a submarine at range of 1,500 to 2,000 yards under normal conditions and, therefore spaced 1,000 yards apart. The radio could transmit the signals to a shore station not more than 15 miles away. By listening to the buoy from which the sound as the loudest an operator could estimate the position of the intruder. The HECP had a Navy underwater detection division to maintain and operate these systems.[86]

German U-boats Patrol the Approaches to Delaware Bay

It soon became apparent, though, that the greatest threat to the region during World War Two would be from German U-boats that lurked just off the Delaware Bay to intercept ships coming out or entering of the bay. In December 1941, just after the U.S. declared war, the first group of new Type IXB U-boats sailed for the U.S. coast in what was called "Operation Paukenschlag", translated as "Drumroll or "Drumbeat". The U.S. Navy was no more prepared and ready for that attack than it had been for the attack on Pearl Harbor. The Commander Eastern Sea Frontier (ESF) was responsible for defense of the coastal waters from Maine to Cape Hatteras. His initial forces contained few ships capable of open ocean patrol or anti-submarine warfare (ASW). Most Navy ships capable of ASW had gone to Great Britain in 1940 as part of the "destroyers for bases" deal and since July 1941 newer destroyers were committed to convoy duty from Newfoundland waters to the mid-Atlantic where the British took over. Thus, during the period January to June 1942, 100 ships were sunk in the ESF area. Of those, 17 were sunk in the Delaware Cape area or approaching that area.[87]

On 13 January 1942, German U-boat attacks officially

Figure 36. Construction of the firing platforms at Battery #5 at Fort Miles, DE on May 21, 1942 before the moving the two 3-inch M1903 guns from Battery Hentig and the two 3-inch M1903 guns from Battery Turnbull, Fort Wadsworth, NY. (RG77 NARA)

Figure 37. US Army Air Corp Dover Air Field was built 4 miles outside of Dover, DE to provide support for observation and bombardment squadrons that would patrol the approaches to the Delaware Bay seeking German U-boats. (Delaware Public Archives 1942)

started against merchant ships along the Eastern Seaboard of North America. From then until early August, these U-boats dominated the waters off the East Coast, sinking fuel tankers and cargo ships with impunity and often within sight of shore due to lack of an effective defense by American forces. In less than seven months, U-boat attacks would destroy 22 percent of the tanker fleet and would sink a total of 233 ships in the Atlantic Ocean and the Gulf of Mexico. These U-boats killed about 5,000 seamen and passengers, more than twice the number of people who perished at Pearl Harbor.[88]

It took six months for the American military to acquire ships and planes, to train personnel to challenge the U-boats, and to develop an effective set of defensive tactics. The US Navy and Army Air Force feuded over who had responsibility for aircraft patrols and bases along the Eastern seacoast. Even into 1943, the zones of greatest danger on the East Coast were near the ports of Boston and New York, the mouths of the Delaware River and the Chesapeake Bay, and off Cape Hatteras. By July 1942, coastal convoys with 40-50 merchant ships received adequate warship escorts and aerial patrols that started to deterred submarines from making attacks. The Dover Army Air Field, 4 miles southeast of Dover, Delaware was assigned to First Air Force which had several units assigned to protect the Delaware Capes from German submarines. These units were the 18th Observation Squadron (65th Obsn Gp), 2 March 1942 – 23 March 1942, 80th Bombardment Squadron (Medium) (45th Bomb Gp), 29 April 1942 – 15 July 1942, and the 39th Bombardment Squadron (Medium)/ 3rd Antisubmarine Squadron (Heavy), 19 July 1942 – 28 February 1943.[89] The Naval Air Station at Cape May, New Jersey operated several fixed-wing Navy scouting squadrons. This air station consisted of a total of 5 paved runways, two blimp mooring circles, parking aprons, hangars, and seaplane ramps. The peak complement of the station was 1,339 personnel which include both Coast Guard and Navy units conducting anti-submarine operations. Blimps from Lakehurst's ZP-12 used the air station at Cape May to patrol the approaches to the Delaware Bay.[90]

The first submarine to arrive in the area off the Delaware Capes after operating further north was *U-123*. The U-boat spent 16-17 January 1942 lingering in the area of Five Fathom Bank on the surface at night looking at the lights of Cape May, Lewes and Rehoboth. Fortunately, it found no targets and moved south. Unfortunately, *U-130* was not far behind. That sub began the campaign by sinking two oil tankers, which were the primary targets during the campaign. The first ships to be lost near Delaware Bay where on 4 and 5 February 1942 when the twin ships, SS *India Arrow* and SS *China Arrow*, both out of Beaumont, Texas and heading for New York with shipments of oil, were torpedoed and sunk. The *India Arrow* was lost off Atlantic City, New Jersey. The *China Arrow* sank off Lewes, Delaware.[91]

February was the worst month and the emphasis remained on oil tankers. The last of the Type IX subs on station *U-103* sunk three more before departing. As the Type IX subs withdrew the Germans threw even more subs into the campaign, using the shorter-endurance Type VII subs crammed with provisions and fuel and supported by supply subs. The first to arrive in the area was *U-578*.

The first warship lost was the USS *Jacob Jones*, a four stacked destroyer, was heading north at a steady 15 knots off the Cape May coast on 28 February 1942 when she was torpedoed. It was the first sinking of a U.S. warship following the Japanese attack on Pearl Harbor. In response to the increasing sinking's, early in the month, ten old destroyers had been released from convoy duty for patrol operations in the Eastern Sea Frontier. Among those was the USS *Jacob Jones* (DD-130). After refit and resupply, she sailed from New York on February 26 with orders to patrol off the coast between the Delaware Bay and Chesapeake Bay. On the morning of 27 February 1942, she encountered the burning wreckage of the tanker SS *Resor*. After unsuccessfully searching for the U-boat, Jones continued south. However, the sub had preceded her and lay in wait. Early morning on 28 February *U-578* fired three or four torpedoes. The first two hit the ship forward, exploding the magazine and destroying the bridge. The forward section of the ship went down with all hands. The next one or two torpedoes hit the stern, causing it to begin

Figure 38. US Navy Naval Station at Cape May, NJ provided support to naval aircraft and blimps as well as patrol craft to guard the approaches to the Delaware Bay. (RG80 NARA 1942)

sinking. Only the mid-section remained afloat and 38 men were able to abandon ship. Then, as the stern sunk, the depth charges exploded killing all but 12 survivors.[92]

Many other ships were lost to enemy fire and torpedoes along the New Jersey and Delaware coastlines during the early months of 1942. The SS *Hvoslep*, a Norwegian freighter, was torpedoed on 10 March just two miles east of the Fenwick Island Shoal buoy. In March, a number of defensive steps were taken that changed the nature of the submarine war. In January, the Civil Air Patrol had formed two Coastal Patrol Bases, at Atlantic City and Rehoboth Beach. On the first of March, they began flying and, on the fifth, a patrol from Rehoboth spotted and dove on its first sub contact, causing the sub, probably *U-94*, to break off an attack on a tanker. In addition, convoys began to be formed in mid-March, to sail in daylight only, with a navy blimp from Cape May scouting ahead. Most importantly, tankers proceeding from the Gulf to Philadelphia were routed into the Chesapeake Bay and north via the Delaware-Chesapeake Canal to Philadelphia (though the canal was closed for six months when vertical lift drawbridge over the canal was destroyed on 28 July 1942, after being struck by the tanker *Franz Klasen*).[93]

As the U-boats moved south looking for tanker targets, there were fewer attacks in the Delaware Capes area, mainly on freighters. In April and May in the ESF area, sinking's declined as the submarines began to operate south of Cape Hatteras in search of tankers. At about 8 PM on Wednesday, 24 June 1942, just as Cape area residents and vacationers were out for the evening and strolling on the Rehoboth boardwalk, an enormous explosion occurred just off the entrance to the Harbor of Refuge. The tug *John R. Williams*, returning to Cape May from a trip to Fenwick and failing to use the mine-swept channel, struck a mine laid by *U-373* on 10 June. The small size of the tug (396 tons) did little to muffle the sound or contain the impact given the large size of the 2,000-pound mine. The sound was heard at a long distance and parts of the tug were strewn over a wide area. Only a lucky 4 of the 18-man crew survived.[94]

Although German submarines no longer routinely operated in Cape waters after 1942, in February 1945, *U-869* a Type XIC, boat heading toward the coast was attacked and sunk by USS *PC-565* well off New Jersey. Another Type XIC boat was also headed there as the war ended. *U-858* had been on patrol for 59 days without having made a single successful attack, when the war ended and it got the word to surrender. The U-boat surfaced in mid-Atlantic near the USS *Muir* (DE-770) and USS *Carter* (DE-112). Those ships turned the U-boat over to the USS *Pope* (DE-134) and USS *Pillsbury* (DE-133) who escorted the *U-858* to Harbor of Refuge where the formal surrender and transfer of its crew to captative was accomplished on the mine pier of Fort Miles.[95]

Permanent Coast Artillery Batteries Come to the Delaware Bay

The reduction in the armament at the forts at the entrance to the Delaware River left only the obsolete, out of service, 12-inch mortars of Batteries Best and Rodney at Fort DuPont, and Battery Arnold's three 12-inch disappearing guns at Fort Mott. On 12 November 1942, the mortars were ordered removed and salvaged as scrap. The press of other activities delayed their removal and the ordnance was still emplaced at the end of 1942.[96]

Meanwhile, at Cape Henlopen the engineers were ready to commence the second of the three 6-inch long range gun batteries. BCN 222 was sited just to the north of the

Figure 39. USS Jacob Jones *(DD-130) was sunk by* U-578 *on February 28, 1942 off the approaches to the Delaware Bay with the loss of 138 sailors. (RG80 NARA 1930s)*

155 mm guns of Temporary Battery No. 22 and the HECP. Construction began on 15 April 1942 and continued until 29 October 1943. Like the other 6-inch battery at Fort Miles, its gun emplacements were founded on pilings driven some 60 feet into the ground. Its design was the

Figure 40. German U-boat cruising on surface as part of Operation Drumbeat were able to dominate the waters off Delaware Bay from January to June 1942. (RG80 NARA 1942)

Figure 41. The sinking of the tanker SS Resor *by the U-578 on February 27, 1942 off the New Jersey coast resulted in the death of 48 men out of its crew of 50. (RG80 NARA 1942)*

same as BCN 221.⁹⁷ On 17 September 1942, while BCNs 221 and 222 were still being built, the War Department directed that they be named in memory of two deceased coast artillery officers, BCN 221 in honor of Lt. Col. Ralph E. Herring and BCN 222 for Col. Charles Hunter.⁹⁸

On 16 April 1942, Colonel Ruhlen relinquished command of the Philadelphia Sub-Sector and the Harbor Defenses of the Delaware in accordance with G.O. No. 8, HQ Philadelphia Sub-Sector and HD of the Delaware. He was temporarily succeeded by Lieutenant Colonel Roscoe, pending the arrival of the new commanding officer, Col. Robert E. Phillips. Plans were then underway to move both the sub-sector and harbor defense HQ to Fort Miles permanently, and advance elements were already at Cape Henlopen when Colonel Phillips arrived.⁹⁹

On 6 June 1942, the Fort DuPont signal station was closed, and on 10 June HQ Philadelphia Sub-Sector, HQ and HQ Battery, HD of the Delaware, and HQ and HQ Battery, 21st C.A., were all temporarily moved to Fort Miles. All that remained at Fort DuPont was the band and a small caretaking detachment from HQ Battery, 21st C.A. A similar detachment was posted at Fort Delaware. The band moved to Fort Miles on 15 July and on 25 July, the temporary change of station of the command elements was made permanent by G.O. No. 7, HQ Philadelphia Sub-Sector.¹⁰⁰

When the construction program was initiated in 1940, an existing railway spur that ran from the main line of the Pennsylvania Railroad onto the Cape Henlopen Military Reservation near the main roadway gate was extended to sites for the future 16-inch batteries. After entering the post, the railway turned south for a short distance, and then eastward at a point several hundred yards directly behind the site for projected BCN 119, the northerly of the two 16-inch batteries. This line was then divided into four separate firing spurs some fifty feet apart.

On 15 March 1942, a train bearing the guns and heavy equipment of Capt. Louis X. Levin's Battery C, 52nd C.A.(Ry) Regiment, arrived at Fort Miles from Fort Hancock with four MkVIM3A2 8-inch guns on M1A1 railway carriages. The battery train included ammunition cars, a plotting room car, and gondola cars containing fifty-foot towers for the battery commander and as base end stations. The latter, however, proved unnecessary, as a number of the permanent concrete towers, although not formally transferred by the engineers, had already been completed and placed in service. Soon after the arrival of the first of the railway batteries, work began on the four emplacements and their service magazines in a wooded area between the Fort Miles parade ground and the Great Dune area, about 2,750 feet from the beach. Each of these

Figure 42. The U-858 *waits in Delaware Bay's Harbor of Refuge on May 14, 1945 as a Coast Guard HNS-1 helicopter and Navy K-Type airship circle above as the German crew is taken to Fort Miles' mine wharf. (RG80 NARA 1945)*

Figure 43. The crew of U-858 *board a US Army truck on the Fort Miles' mine wharf on May 14, 1945 to be transported to Fort Miles' brig. (DNREC 1945)*

Figure 44. Protective 8-inch railway gun emplacement under construction at Fort Miles, DE in July 1942 using earth and paper cement bags. (RG77 NARA 1942)

Figure 45. 8-inch MkVI M3A2 gun on M1A1 railway carriage in firing position with outriggers deployed. (NARA 1942)

four emplacements consisted of a horseshoe shaped earthen epaulement built around the end of each firing spur. These eleven-foot revetments were made quasi-permanent, with an interior lining of paper cement bags laid in lieu of less permanent burlap sand bags. The battery's plotting room car was placed on a siding about 250 yards to the rear of the battery. The men of Battery C, 52nd C.A., lived under canvas adjacent to the battery site. By 25 April their guns were emplaced and ready for service.[101]

This railway artillery was augmented on 10 September 10, 1942, when HQ and HQ Battery, 2nd Bn., and Battery D, 52nd C.A., under Lt. Col. Thomas B. McConnell, CAC, arrived from Fort Hancock. The post railway was extended farther to the south and then turned eastward toward the beach. There, firing spurs were laid for four more 8-inch railway guns. Battery D's guns were emplaced about halfway between Battery Smith and BCN 519, south of the Great Dune, about 1,250 feet from the ocean and about 4,500 feet southeast of Battery C's position. Preparation of the second railway artillery site advanced rapidly, and by November 9, 1942, all four of Battery D's 8-inch guns were ready for service in firing positions of basically the same design and construction as those for Battery C. The guns manned by Battery C were designated as Tactical Battery No. 20, while those manned by Battery D were designated Battery 21.[102]

As the buildup at the capes continued, more of the armament was relocated seaward. In January 1941, the casemating of Batteries Hall and Haslet had briefly been a part of the modernization program for the harbor defenses. In May 1941, however, all preparations for casemating Fort Saulsbury's guns were halted and a plan was developed to move Battery Hall's 12-inch guns to Cape Henlopen. There they were to be temporarily emplaced in open barbette emplacements near the south end of the reservation, pending the completion of the two projected

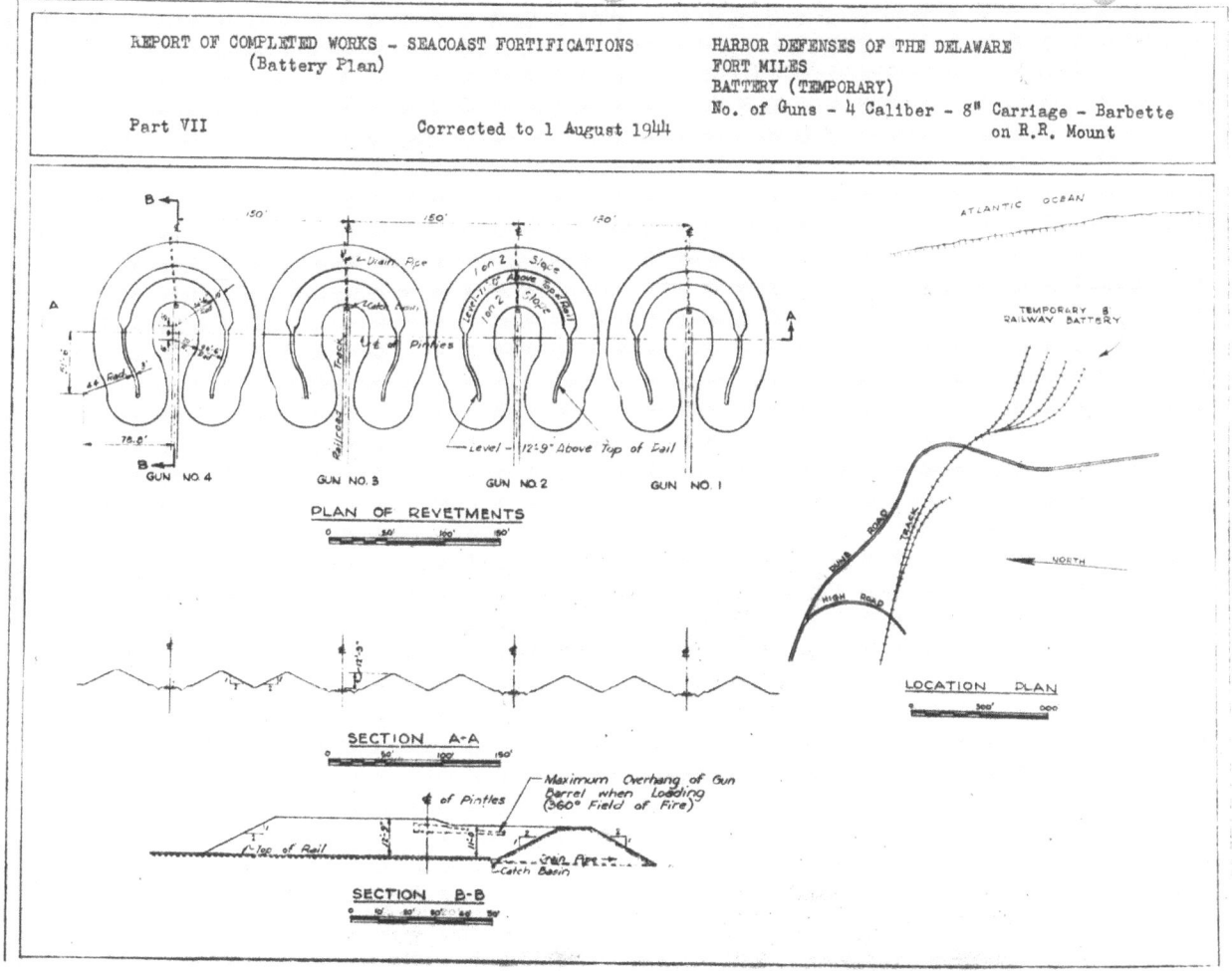

Figure 46. RCW Form 7 for Battery #21 showing the plan and sections for the revetments and location plan at Fort Miles, DE for the four 8-inch MkVI railway guns. (RG77 NARA 1944)

Figure 47. Battery #20 (four 8-inch MkVI railway guns) with protective revetments and camouflage (snow fencing) with the buildings of Fort Miles's main cantonment in the background (note the tall radar tower). (DNREC 1944)

batteries of casemated 16-inch guns. Once the two 16-inch batteries were completed, the 12-inch guns would be withdrawn if no longer required. On 27 July 1942, authorization was granted to move Battery Hall's guns to Fort Miles. In the meantime, construction began on the two concrete gun blocks to temporarily accommodate the M1895M1 12-inch guns and their barbette carriages.

Meanwhile, the War Department contemplated sending the 12-inch guns of Battery Haslet, in maintenance status at Fort Saulsbury, to Trinidad, British West Indies. Final action on this transfer was withheld, and ultimately was canceled. When the decision was made on 13 November 1942, to suspend the construction of BCN 119, construction of the incomplete gun blocks for the 12-inch guns was also halted.[103] On 8 January 1943, the decision was made to emplace Battery Haslet's guns and carriages in BCN 519, instead of Battery Hall's. At this point, the guns of both batteries were still at Fort Saulsbury. Battery Haslet's armament was not actually moved to Fort Miles until the spring of 1943. Battery Hall's guns remained at Fort Saulsbury in maintenance status.[104]

In the 1940 modernization program, the Fortifications Board had planned for two batteries of 16-inch guns at Cape Henlopen. The first of these, BCN 118 (Battery Smith), was completed on 31 October 1942, with delivery of its armament scheduled for November. Although authorized in 1934 as one of the initial casemated 16-inch batteries, Battery Smith was not actually begun until March 1941. When construction finally began, it was built in accordance with the designs for the 1934 program, which called for the casemates to be splayed outward. In the rear of each casemate was a 10-foot by 22-foot

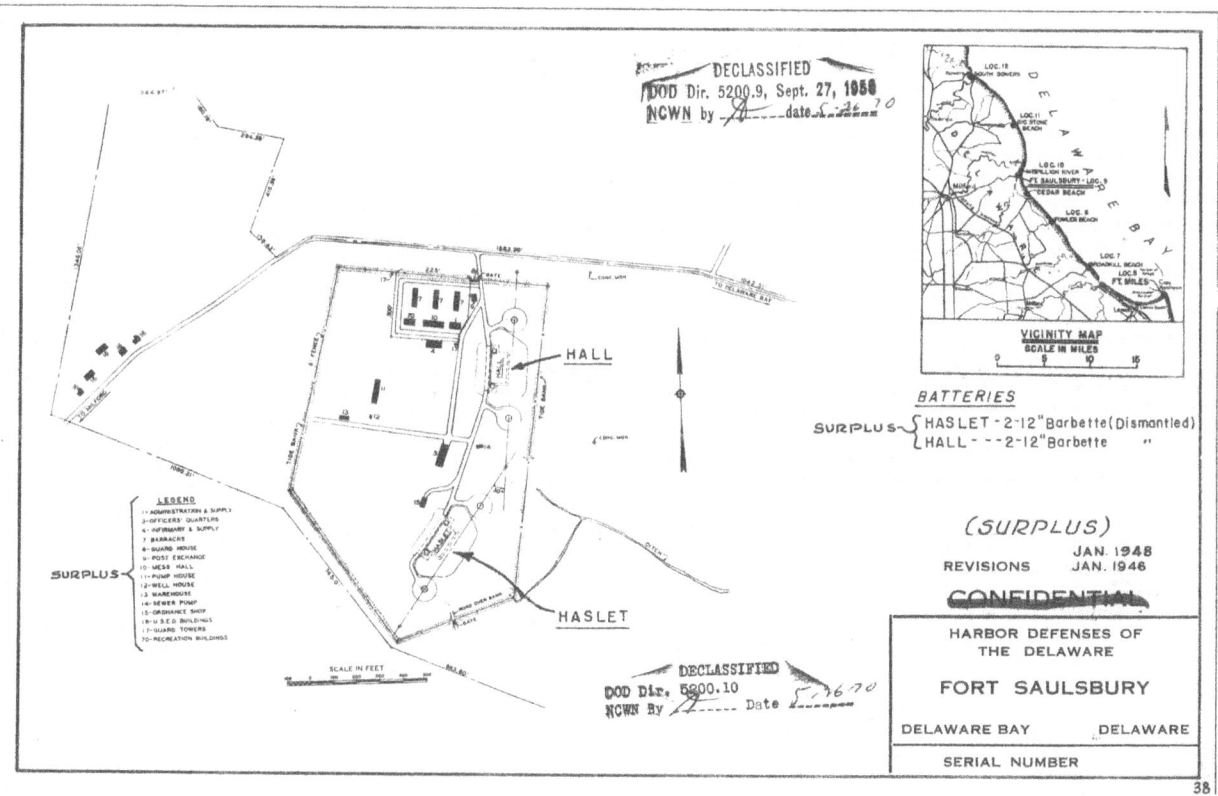

Figure 48. Final site plan for Fort Saulsbury in January 1948 shows the new WW2 support buildings while guns of Batteries Hall and Haslet have been dismantled and fort is now surplus of US Army needs. (NARA 1948)

storeroom. Each casemate was connected to the front corridor of the central magazine and service traverse by an angled ammunition service passage. Along the rear wall of the corridor were two projectile rooms, 40 feet wide and 18 feet deep; and two power magazines, 50 feet wide and 25 feet deep. At the end of the shell room for the Number 2 gun there were two storerooms, each 10 by 15 feet, while at the opposite end of the corridor was a 32-foot by

Figure 49. The only known photograph of completed Battery Smith with its 16-inch MkII gun in place (the lack of images is due to wartime security regulations barring photos of the coast defense installations). (DNREC 1945)

10-foot latrine. Midway along the front corridor, a central corridor perpendicular to the front corridor provided access to the power plant located at the rear of the structure. This contained, in addition to the 50-foot long and 25-foot wide room for the generating equipment, two 10-foot by 12-foot storerooms, a 15-foot by 25-foot air compressor room, and a 15-foot by 10-foot water cooler room. Along the rear was the muffler gallery.[105]

While the battery structure was constructed in accordance with the standardized 1934 plans, the gun carriages of BCN 118 contained some improvements over the carriages installed in the late 1930s. The two 16-inch MkII MI guns of the battery were to be mounted on M4 barbette carriages. These differed only slightly from the M3 carriages—the gun was provided with a three piece, 200,000-pound shield of cast steel armor that was attached to the carriage. The armor plating of the two guns was four inches thick, twice that of the batteries constructed under the 1934 authorization.

It had also been discovered that the cyclic rate of fire for the 16-inch guns installed in the prototype batteries of the 1934 program was limited to a round every 90 to 150 seconds, due in large part to the underpowered traversing

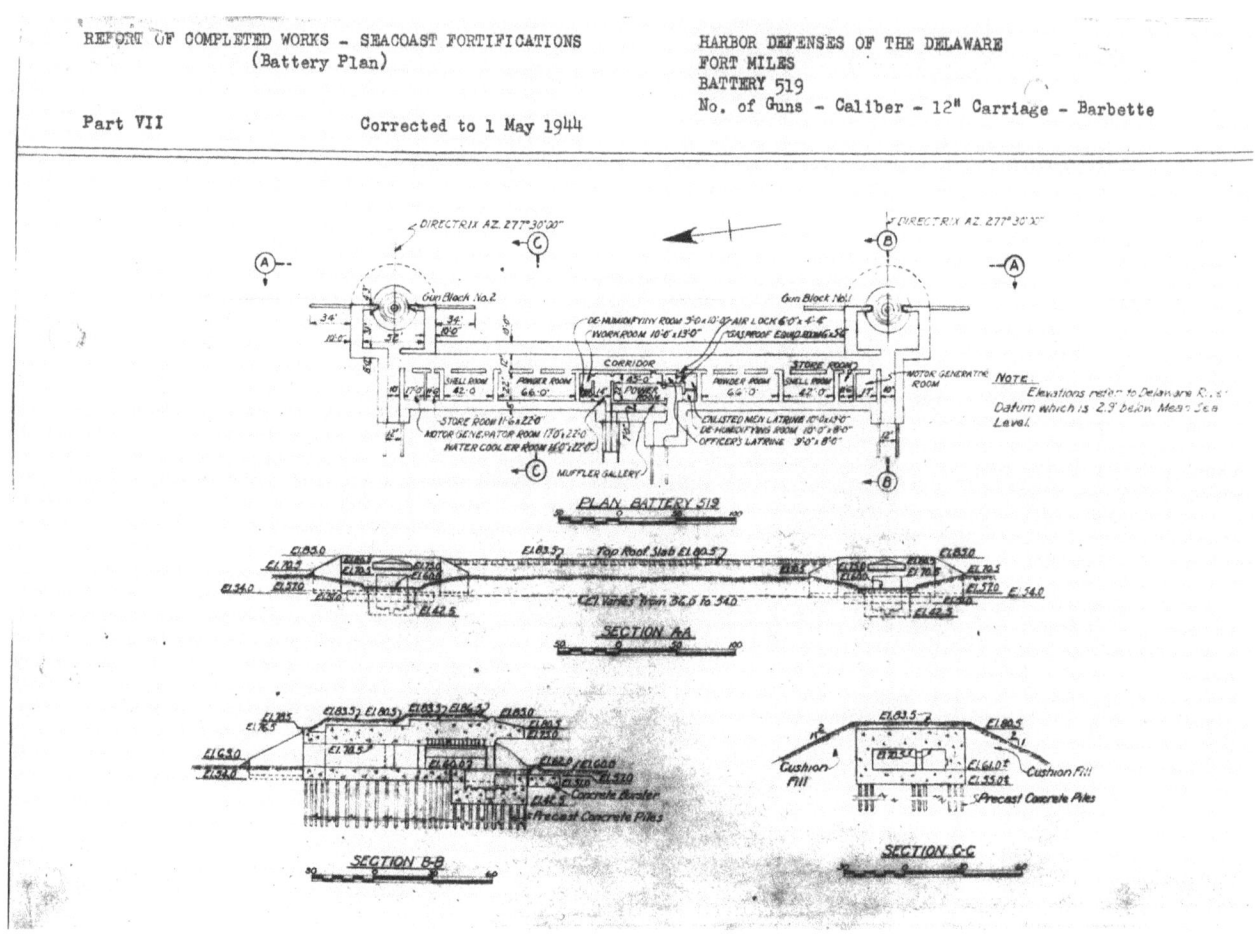

Figure 50. RCW Form 7 for Battery #519 showing the plan and sections for the casemated battery (which replaced the planned 16-inch casemated battery #119) at Fort Miles, DE for two 12-inch M1895M1A4 guns. (RG77 NARA 1944)

and elevating motors. Warships mounting 16- inch guns could maintain a rate of fire of one round every twenty seconds, four to five times the rate of the early 16-inch casemated batteries. To increase the rate of fire, and because of the additional weight of the improved shield, it was necessary to increase the horsepower of the traversing motor and substitute all-electric drives to supplement or replace the hydraulic traversing and elevation drives. A 440-480 volt AC power plant consisting of three 15 KW Westinghouse generators was installed in the battery power room. This was sufficient to enhance the traversing as well as the elevating speed, enabling the gun to be placed in the loading position more rapidly, increasing the cyclic rate of fire to three rounds per minute.[106]

The second battery, BCN 119, authorized in 1940, was to have been built several hundred yards north of Battery 118. Initially, it was given a high construction priority of number 14 on September 1940, and $625,000 was allotted to initiate construction. Some excavation was begun in 1941, but actual construction had not begun on 11 August 1941, when the modernization program was again reviewed; BCN 119's priority was dropped to 31, deferring construction. In November 1941, construction was suspended. Two days later the battery was eliminated from the harbor defense project altogether, and replaced by BCN 519.[107] The decision to alter the armament of the battery also resulted in the modification of the battery plans to accommodate the 12-inch guns. The design of battery differed somewhat from the plans of Battery Smith. The main corridor of Battery 519 was straight between the gun casemate. In addition, the power room was directly off the main corridor in Battery 519 rather than at the end of a short passageway. Construction began 15 November 1942.[108] Emplacing the M1895M1A4 12-inch guns on their M1917 barbette carriages began in March 1943, and the battery was complete by 13 August 1943. As with most of the batteries and other installations at Fort Miles, Battery 519 was manned before it was formally transferred to the coast artillery on 18 February 1944. The plotting and switchboard room for Battery 119

Figure 51. The service corridor with an overhead rail system to move shells to the 12-inch gun casemates (this image is from Battery #520 on Sullivan's Island, SC, as a sister battery of Battery #519). (RG111 NARA 1943)

Figure 52. The 12-inch M1895M1A4 on M1917 barbette carriage in gun casemate without shield (this image is from Battery #520 on Sullivan's Island, SC, as sister battery of Battery #519). (R111 NARA 1943)

Figure 53. The power generation room with three 240HP diesel powered motor (Buckeye) 5KW generators (Westinghouse) with supporting air compressor (this image is from Battery #520 on Sullivan's Island, SC, as sister battery of Battery #519). (R111 NARA 1943)

was completed and transferred to the coast artillery on 21 December 1943. This gas-proofed, bombproof structure was placed in service as the temporary HDCP as soon as it was completed. As the HDCP, it also housed the post's protected switchboard for Fort Miles.[109]

Manning the Coast Defenses of the Delaware Bay

At the end of the first year of the war, the organization of the HD of the Delaware was well developed. Much of the temporary armament was in place, pending completion of the modern batteries of the permanent project. Construction of these modern batteries, now consisting of one battery of 16-inch guns, one battery of 12-inch guns, three batteries of 6- inch guns, and three batteries of 90 mm guns, was advancing. As the war entered its second year, the temporary harbor defenses were composed of gun groups and a mine command at Fort Miles, and a separate battery at Cape May. This tactical organization remained in effect until 1944 with only minor modifications.

Gun Group 1 was made up of the two 8-inch railway batteries (Tactical Batteries 20 and 21) at Fort Miles, manned by Batteries C and D, 2nd Bn., 52nd C.A. On May 1, 1943, the 2nd Bn., 52nd C.A. was redesignated the 267th C.A.(Ry) Bn. (Sep).[110]

Gun Group 2 had the four 3-inch rapid-fire guns of Battery No. 5, manned by Battery C, 261st C.A. Bn.; the four 155 mm GPF guns of Battery 22, manned by Battery A, 261st C.A. Bn.; and the two mobile 90 mm guns of AMTB Battery 5A manned by a detachment of Battery C of the 261st.

The Mine Command was operated by Batteries A and B, 21st C.A., while the separate battery at Cape May, Battery

25, was manned by Battery C, also of the 21st. During 1943, the 155 mm guns of Battery 25 at Cape May were grouped with the mines to form **Gun Group 3**. The remaining 90 mm guns of Battery 5A and Battery 5B were added to Gun Group 2 during 1943. These 90 mm guns were delivered piecemeal beginning in November 1942. The last of these dual-purpose weapons were not in place until May of 1943.[111]

While efforts to activate a second battalion of the 21st C.A. Regiment were made in 1942, this expansion did not occur, and the regiment was ultimately redesignated a separate battalion with 23 officers, three warrant officers, and 551 enlisted men. The 261st C.A. Bn. had 30 officers, three warrant officers, and 576 enlisted men. The HQ and HQ Battery of the harbor defenses and the Philadelphia Sub-Sector had eight officers and 33 enlisted men assigned. In addition, a provisional harbor defense training and replacement battery was operated by the harbor defenses with some 273 enlisted personnel on hand on 1 January 1943. The 12th Mine Planter Battery was composed of one officer, six warrant officers, and 42 enlisted men.

Noting that the Axis forces had been driven back on the high seas and seeing the increasing magnitude of operations in Europe and Asia, the Eastern Defense Command on 29 October 1942 noted that an attack by a major axis fleet was deemed a remote probability and actions of the "commando type" were deemed improbable and the only likely hostile action was isolated raids by submarines and light vessels. Consequently, the War Department directed that the category of defense "B" be reduced to category "C" for batteries in excess of 6-inch, other than the 155 mm batteries. Batteries of 6-inch or less were to be retained in category "B".[112]

In addition to not authorizing additional personnel to activate the 2nd and 3rd battalions of the 21st C.A., the EDC redesignated the 21st and the 261st as "limited-service units." To release men for overseas duty, the War Department on 11 November 1942, authorized a temporary 15 percent overstrength for the 21st and 261st C.A. Bns. to replace general service personnel with limited service personnel. When the training of the limited service replacements was completed, the general service troops were to be reassigned to the Army Ground Forces within sixty days. The 2nd Bn., 52nd C.A., was maintained as a general service organization.[113]

At the close of 1942, the report of the S.1 of the

Figure 54. The only known photograph of completed Battery #519 with its 12-inch M1895M1A4 gun with shield in place (the lack of images is due to wartime security regulations barring photos of the coast defense installations). (DNREC 1945)

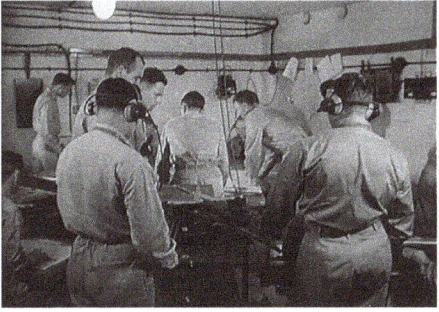

Figure 55. The wartime training in the plotting room receiving targeting data from assigned fire control towers and by using mechanical computers and plotting boards translate this information into firing data to be transmitted to the battery's guns (this image is from Battery Hall at Fort Saulsbury, DE in 1942). (NARA 1942)

Philadelphia Subsector and Harbor Defense of the Delaware showed a total strength of 135 officers, 13 warrant officers, and 2,932 enlisted men in the command. The forces of subsector remained fairly stable at this figure during the time as it was fully operational or until the Battalion Combat Team made up of the 2nd Battalion 113th Infantry and Battery C of the 165th Field Artillery, based in Georgetown, DE was removed from its command. Of these totals, 85 officers, 13 warrant officers, and 1,796 enlisted men were in the Harbor Defenses, while 50 officers, no warrant officers, and 1,136 enlisted men were in the Battalion Combat Team.[114] With the new year came another reorganization to the Philadelphia Subsector. The Miles Groupment was eliminated from the command due to the compact nature of the defense armament now located entirely on the capes. The Infantry Combat Team was still in the subsector's line of command, that mobile combat team not disappearing until 1 November 1943. Otherwise, basically, this organization of the subsector command held until 1944.

The first phase cantonment buildings at Fort Miles occurred between August 1941 and March 1942. Although these buildings no longer exist, they illustrate the variety of plans and materials used at Fort Miles during the course of cantonment construction. These barracks, supply and administrative buildings, recreation buildings, and mess

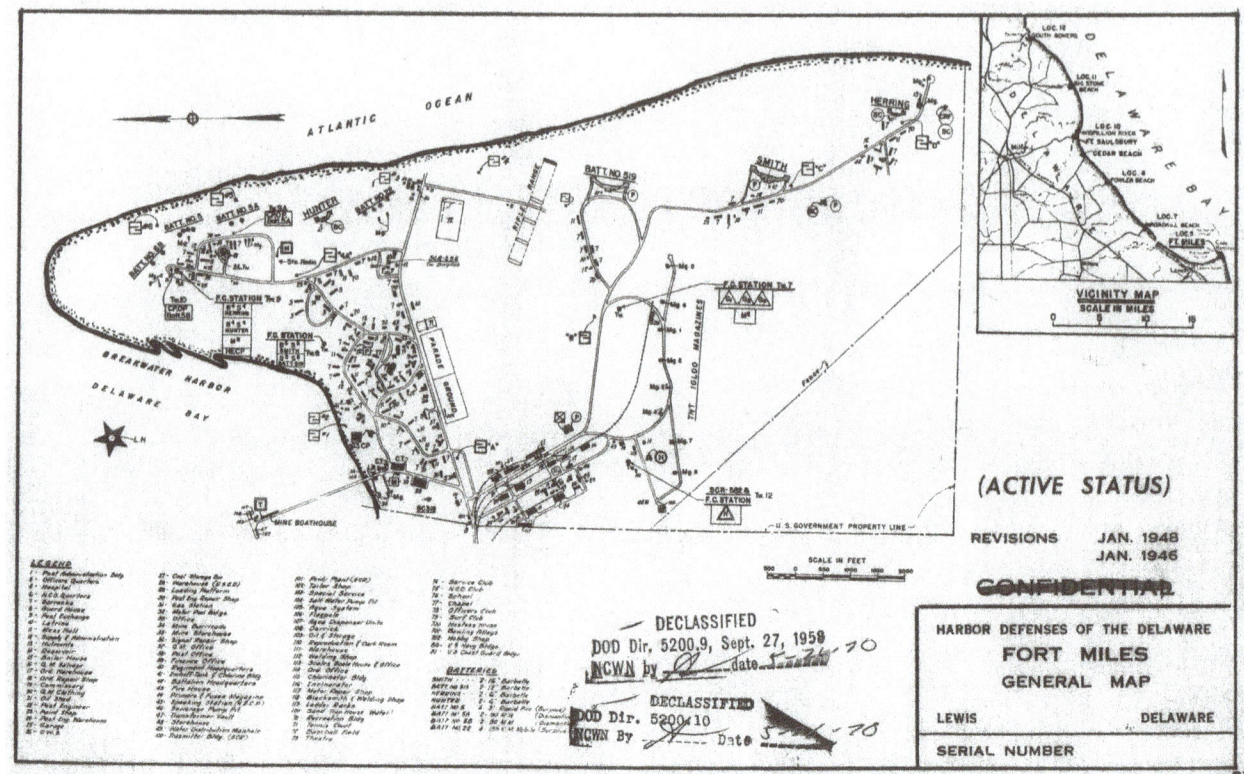

Figure 56. The general map of Fort Miles, DE showing the numerous structures built between 1941 and 1944 to support the fort's coast defense role as only a few buildings existed at Cape Henlopen before Fort Miles was established and by the date of this map in 1948 there is a city of building. (RG177 NARA 1948)

halls were spurred by the need to house, supply, and feed the first troops at Fort Miles. These men operated and oversaw the fort's earliest defensive structures: 155 mm GPF guns and the controlled mine defense. In a letter titled "Temporary Construction at Cape Henlopen, Delaware," the Adjutant General requested that the Chief of Engineers begin construction on temporary mobilization-type housing for 551 enlisted men and 24 officers. He ordered that "only the minimum facilities and utilities consistent with health, sanitation, and recreation be constructed."[115]

A November 1941 plan shows thirty buildings located around the triangle formed by Post Lane, Officers Road, and Point Road. Although no series numbers are given, the plan includes acronyms for each of the buildings. On the basis of correspondence issued by the Post Engineer, all of the buildings are identifiable. Barrack and mess hall capacity, identified in the letter and table by the numbers following BKS and M, as well as the absence of separately housed latrines point to the use of either a 700 series plan or an 800 series plan for this initial cantonment construction. Furthermore, the use of a 700 series or 800 series plan also corresponds to the Adjutant General's order to construct "mobilization" housing. Termed "mobilization" buildings, the 700 and 800 series plans were used during this time period, each presented a plan for barracks to house 63 enlisted men, and both included lavatories within the floor plans.[116]

The demolition of the original cantonment construction

Figure 57. Constructed in late 1943, this cement block and wooden roofed building (Bldg. T600 Company Day Room) represents the last phase of cantonment construction at Fort Miles, DE. (McGovern Collection 2015)

Figure 58. Standard 90mm Anti-Motor Torpedo Boat (AMTB) batteries had two fixed mount guns and two mobile carriage guns along with two 37mm guns as seen in this image of an AMTB at Fort Story, VA. (NARA 1945)

when Fort Miles became a state park and the lack of primary sources that stipulate the type of plan, however, made it difficult to determine which standardized series the Post Engineer used. Moreover, 63-Man barracks for both the 700 Series and 800 Series plans exhibit similar dimensions: 29.5 feet by 80 feet. Two facts are certain, however. The reference in the Area Engineer's letter to the 172-Man mess hall indicates the use of an 800 Series plan, as does a period photo illustrating a Fort Miles barracks with enclosed brick chimney stacks which is only used for the 800 Series. Therefore, it is likely that the first phase of cantonment construction at Fort Miles utilized Series 800 plans.[117]

The second phase of cantonment construction most likely occurred between the winter of 1942 and summer of 1943. Based on the modified Theater of Operations plans, these one-story buildings measured twenty by one hundred feet, were constructed with concrete block walls, were built on concrete slab foundations, and were fitted with wood framed roofs. Over eighty modified Theater of Operations buildings were built during this time period. The site plans included a combination of barracks, latrines, mess halls, supply and administration buildings, and recreation buildings based on housing capacity and division ratios. Although a large portion of this Theater of Operations cantonment construction lay just west of the original cantonment (the 800 Series buildings), small encampments located near the fort's Naval Meteorological

Figure 59. The 90mm M1 gun on M2 mobile carriage was used (along with the M1A1 mobile carriage) and the 90mm T3/M3 fixed mounts for the three AMTB batteries (Batteries #5A and #5B at Fort Miles, DE and Battery #7 at Cape May, NJ). This restored 90mm gun is part of the Fort Miles Museum collection at Fort Miles, DE. (McGovern Collection 2019)

Station, Battery Smith, and Battery Herring were also built during the same time period.[118]

The modified Theater of Operations buildings relied on plans distributed from the Corps of Engineers Office in Washington, D.C. A revision of the Army's cheapest temporary building, the modified Theater of Operations housing was given just enough "structural stability...to meet the needs of the service which the structure [was] intended to fulfill during the period of its contemplated war use." Lumber stud and rafter construction, tar paper and batten walls with 15 pound-felt sheathing, and the absence of an interior lavatory characterized the most of the one-story modified Theater of Operations Buildings built elsewhere. An important aspect of the Theater of Operations construction at Fort Miles was the use of concrete block, rather than the specified wood frame plan.[119]

The last phase of cantonment construction occurred between July 1943 and March 1944. Four barracks, one recreation building, one supply and administration building, one latrine, and one mess hall were placed along the street, Sandy Lane, that led to the last of Fort Miles' batteries, Battery 519. Although they share similar measurements to the barracks in the main cantonment, twenty by one hundred feet, the buildings near Battery 519 exhibit slight design modifications. Side-gabled, the one-story concrete block buildings contain thirteen openings on both their front and rear elevations. Their front elevations are distinguished by an asymmetrical pattern of window and door openings. Doors are located in the second, seventh, eighth, and twelfth bays. In addition, these barracks concrete block chimney stacks on their front elevation and rear elevations.[120]

Within three years, the Area Engineer used three different plans at Fort Miles. Although the initial building phase used the materials called for by the 800 series plans specifications, both the modified Theater of Operations in the main cantonment and the T.O. 700 series utilized

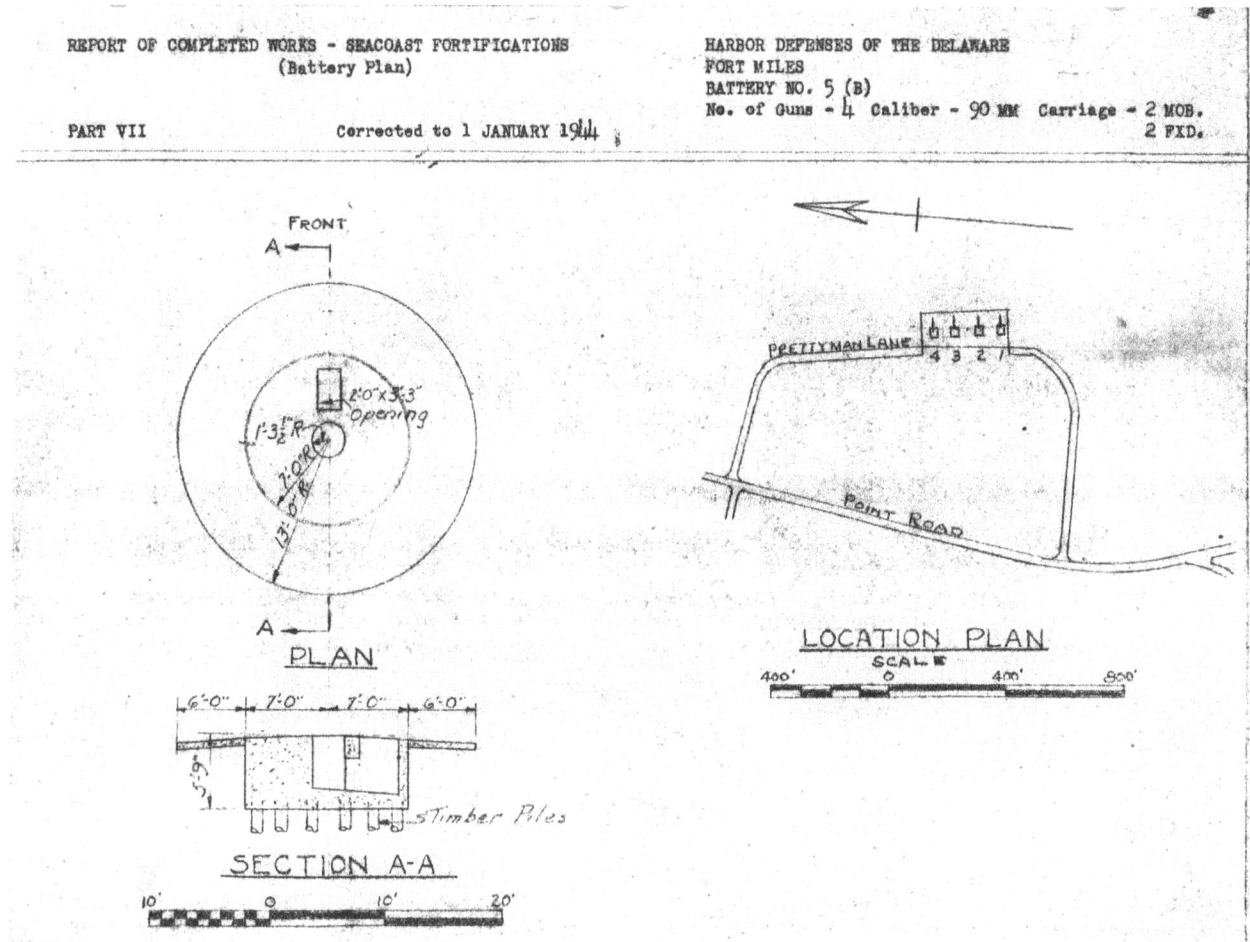

Figure 60. RCW Form 7 for Battery #5B showing the plan and sections as well as the location plan for the AMTB battery at Fort Miles, DE. (RG77 NARA 1944)

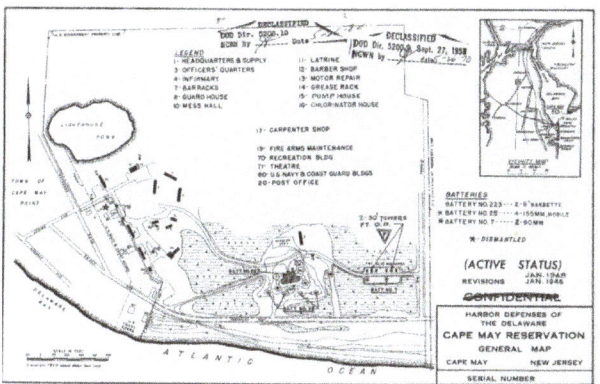

Figure 61. General map of defenses of Cape May, NJ with Battery #223 (two 6-inch shielded guns), Battery #7 (four 90mm guns), and the temporary Battery #25 (four 155mm GPF on Panama mounts) in January 1948. (RG177 NARA 1948)

Figure 62. The final Harbor Entrance Control Post (HECP) was a wooden two-level structure located on top of Fire Control Tower #9 which is located near the tip (it was in 1943 but not now) of Cape Henlopen at Fort Miles, DE. (RG77 NARA 1943)

concrete block in their construction. Fort Miles contained a rare adaptation of standardized plans, it appeared that four factors may have contributed to the use of this particular material: the autonomy of the Area Engineer, the site's environment and topography, the commercial and industrial importance of the Delaware River and Bay, and the material and labor shortage of 1942 – a date that corresponds to the construction of the first concrete block barracks at Fort Miles. The use of concrete block in the construction of the garrison building have allowed a good sample of these "temporary" buildings to survive more than 80 years.[121]

The daily life of the troops assigned to the defenses of Delaware Bay consisted of drills, alerts, physical training, guard duties, instructional classes, kitchen and general maintenance of both the garrison area and tactical defenses. Weather at the Capes varied from hot summer to cold winters, while at Fort Miles the smell of the fish packing plant at the nearby Lewes Beach was an everyday feature. The movement from winterized tents to barrack buildings with hot water was welcome development for the troops assigned to Fort Miles and to the sub-post at the Cape May Military Reservation. The first of the new Tactical Operations buildings constructed were messes with kitchens as food (both qualities and quality) was very important factor in the soldier's lives. Holiday's called for special meals and celebrations. Along with new messes, a permanent post exchange (PX), which sold toilet articles, ice cream and soda, and beer (restricted to 3.2% in alcohol content), and a post infirmary (with a 10-person capacity) provided medical and dental services were constructed.

Keeping the about 2,500 soldiers fully occupied involved social activities as well. Both a gymnasium and theater were constructed on base. Basketball, baseball, and football teams were important to the morale. These teams would play other military teams in the area and would also draw support from the surrounding communities. A variety of officer and enlisted clubs were established along with recreation buildings to house them. Theater and music groups were formed and rotating entertainment was provided. The USO sponsored dances with national known big bands and visit by celebrities occurred. The local Red Cross provided support, especially local women for on-base dances, which were important events for soldiers. The various units stationed in the area produced newsletters that covered social and athletic matters, as we as personnel transfers and leaves. A chapel was built within the base to allow for religious services, while troops were also allowed leave to attend churches in nearby towns. As the 261st C.A. Bn. was formally a national guard unit, many of its men were from the local area and had wives or girlfriends nearby, so gaining leave was an important event for these men. Local towns had movie theaters and USO clubs for service men to visit while on leave. One of the unique social activities for the troops was fishing and swimming due to their location on Delaware Bay and Atlantic Ocean.

The first of the 90 mm M1 guns and their M1A1 mobile carriages arrived in the harbor defenses in November 1942, and on 1 January 1943, construction of two anti-motor torpedo boat batteries at Fort Miles and one at Cape May was begun. The two batteries at Fort Miles, sited on Cape Henlopen on either side of Battery 5, were designated Batteries 5A and 5B. Upon completion of these two AMTB batteries the 3-inch guns of Battery 5 were to be inactivated. The AMTB battery at Cape May was designated Tactical Battery No. 7, and would, upon completion, replace the 155 mm guns of Temporary Battery No. 25. (The former Battery No. 7 at Reedy Island was inactivated in January 1943.[122]

On 26 February 1943, three of the four mobile emplacements of the two AMTB batteries on Cape Henlopen were turned over to Battery C, 261st C.A. Bn. On 10 March with the temporary battery at Reedy Island now inactivated and its armament in storage, the detachment of Battery C of the 261st that had been manning that battery was reassigned to the 90 mm guns of the Fort Miles AMTB batteries. One of the mobile guns of AMTB Battery 5A and both of AMTB Battery 5B were manned on 27 March. The remainder of Battery C continued to man the 3-inch guns of Battery 5 until 23 May 1943, when Battery 5 was placed in readiness class "B," and the men of Battery C assigned exclusively to the 90 mm guns. It was not until the end of May, however, that the six fixed T-3 (M3) mounts and their M1 90 mm guns were received. By 15 June the fixed carriages for Battery 5A and 5B were emplaced and ready for service. Although manned and in service, the calibration firing of Batteries 5A and 5B was not carried out until 11 and 18 November respectively. The two batteries were not formally transferred to Battery C until 21 December 1943.[123] The AMTB also served in a dual-purpose role providing anti-aircraft defense for the coast defenses. Supporting the 90 mm guns were fourteen 37 mm or 40 mm guns of a Bofor design. Additionally, fourteen anti-aircraft 50 caliber guns were emplaced at both Fort Miles and Cape May.

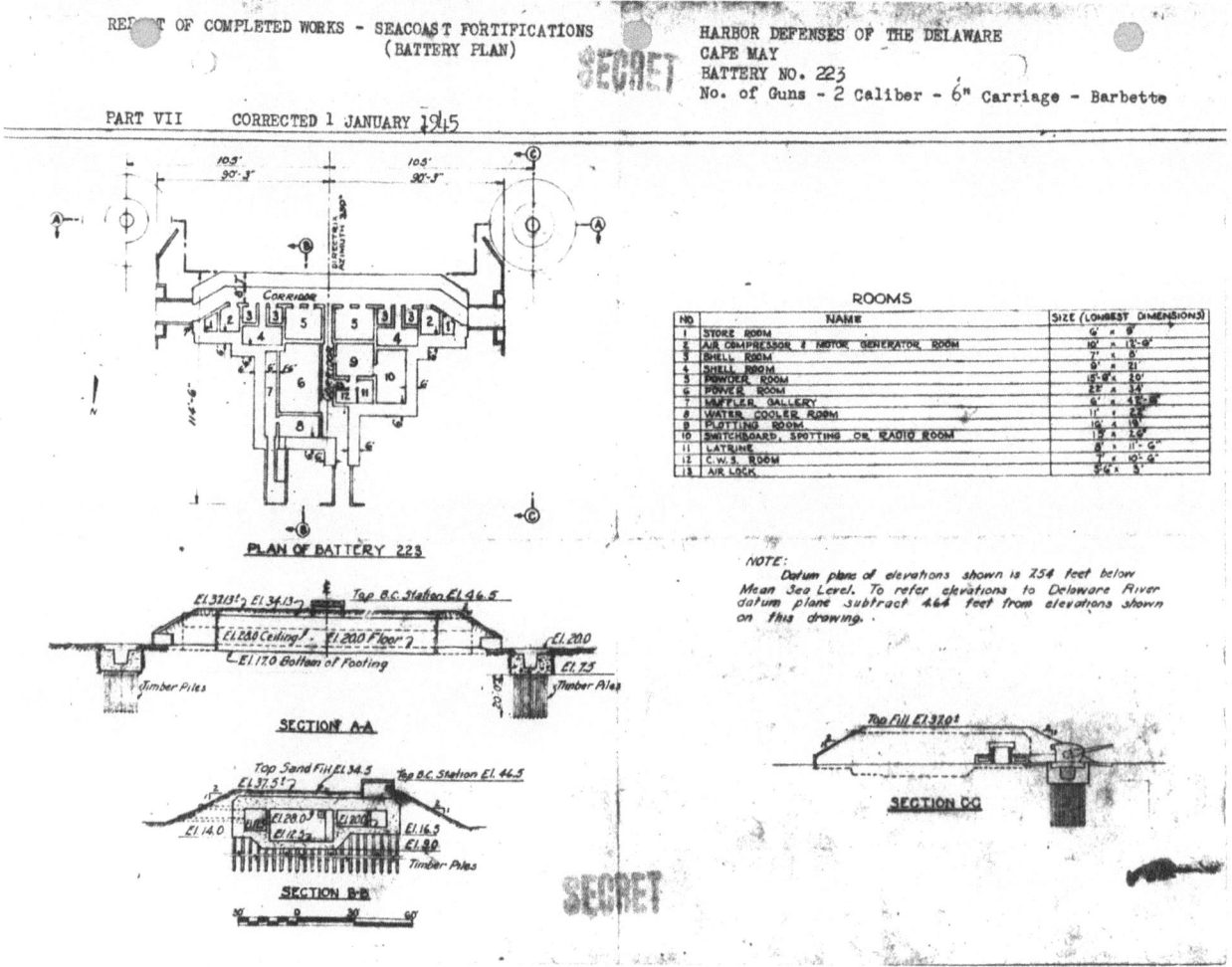

Figure 63. RCW Form 7 for Battery #223 showing the plan and sections for the Series #200 6-inch battery at Cape May Military Reservation, NJ. (RG77 NARA 1945)

Figure 64. The 6-inch M1903A2 gun and shielded barbette carriage M1 of Battery Hunter (#222) at Fort Miles, DE (note the wooden fencing to keep drifting sand from burying the gun). (DNREC 1944)

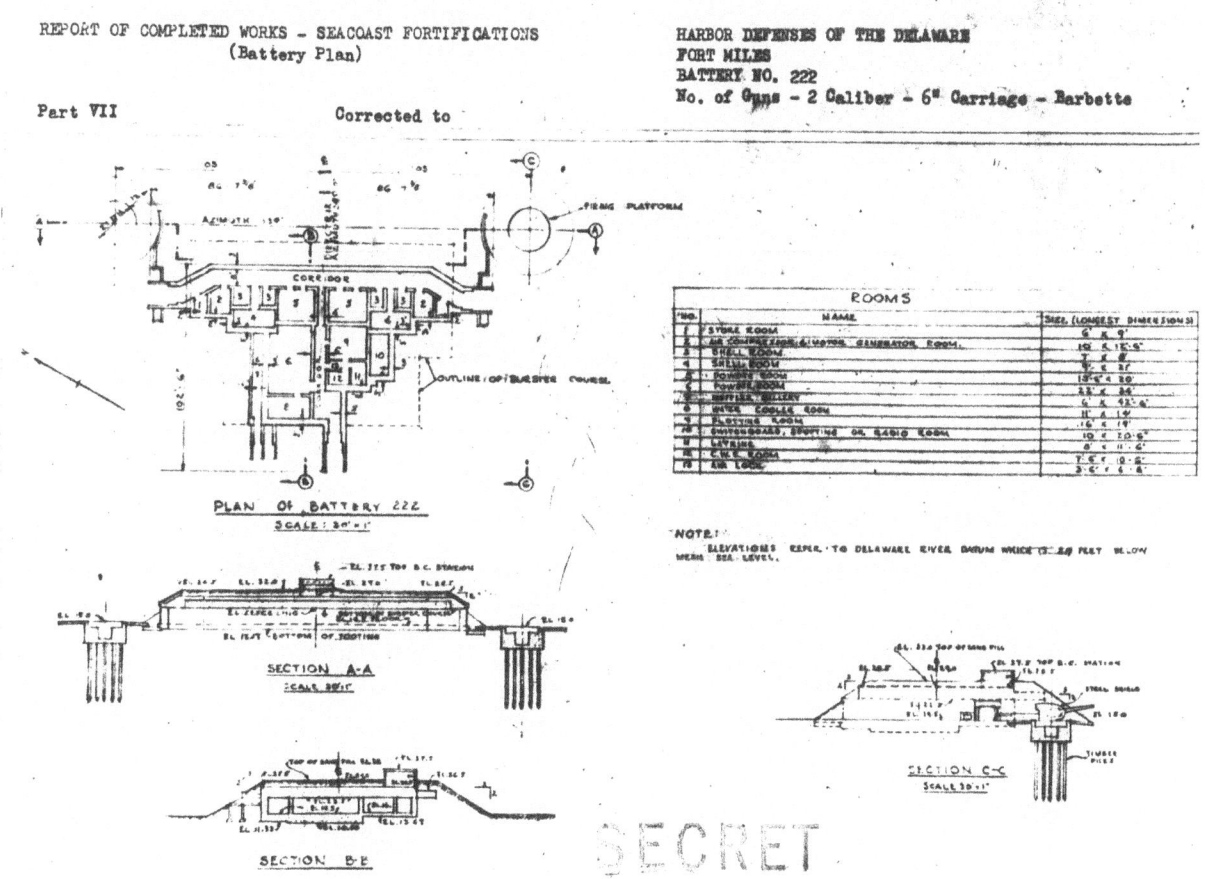

Figure 65. RCW Form 7 for Battery #222 (Hunter) showing the plan and sections for the Series #200 6-inch battery at Fort Miles, DE. (RG77 NARA 1945)

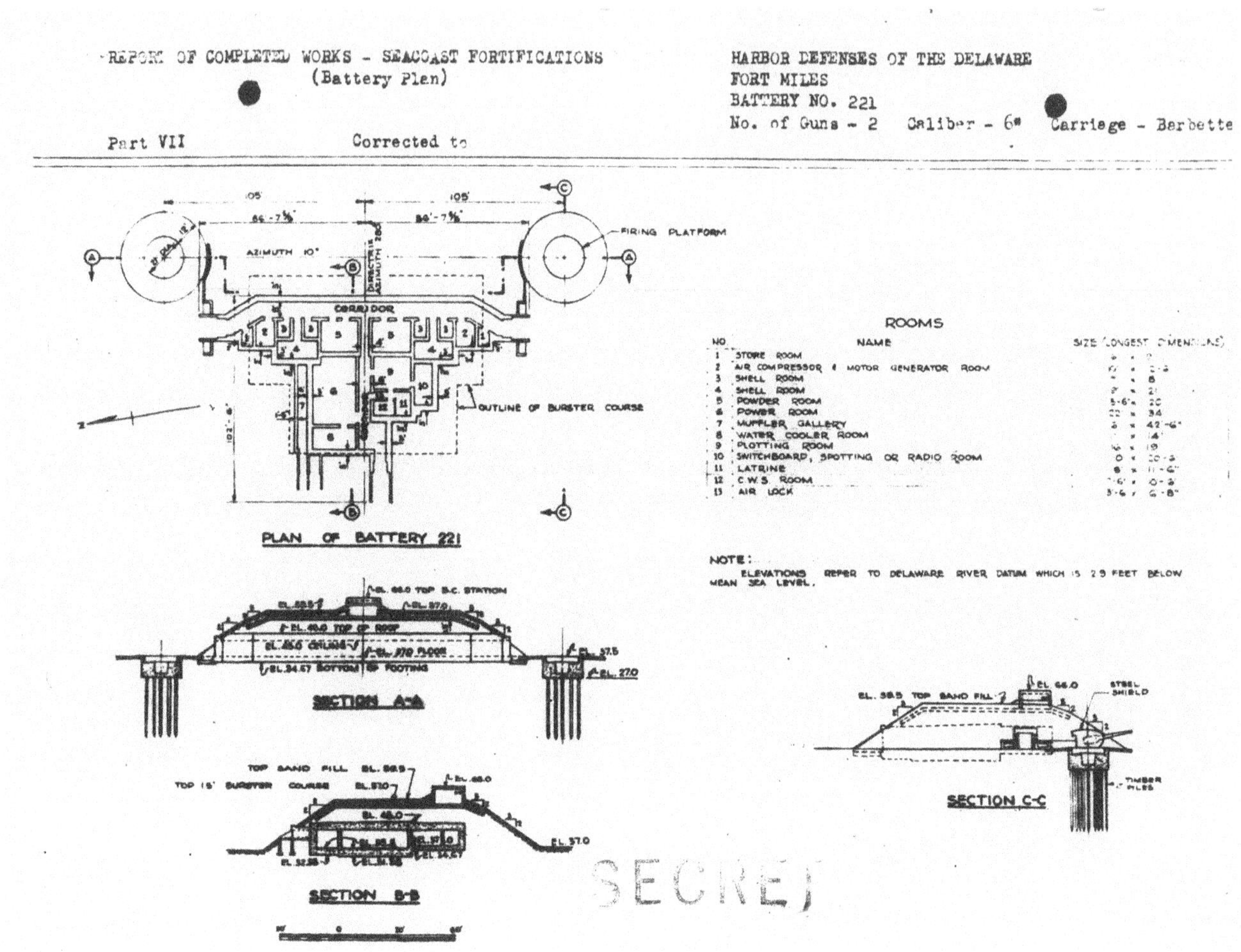

Figure 66. RCW Form 7 for Battery #221 (Herring) showing the plan and sections for the Series #200 6-inch battery at Fort Miles, DE. (RG77 NARA 1945)

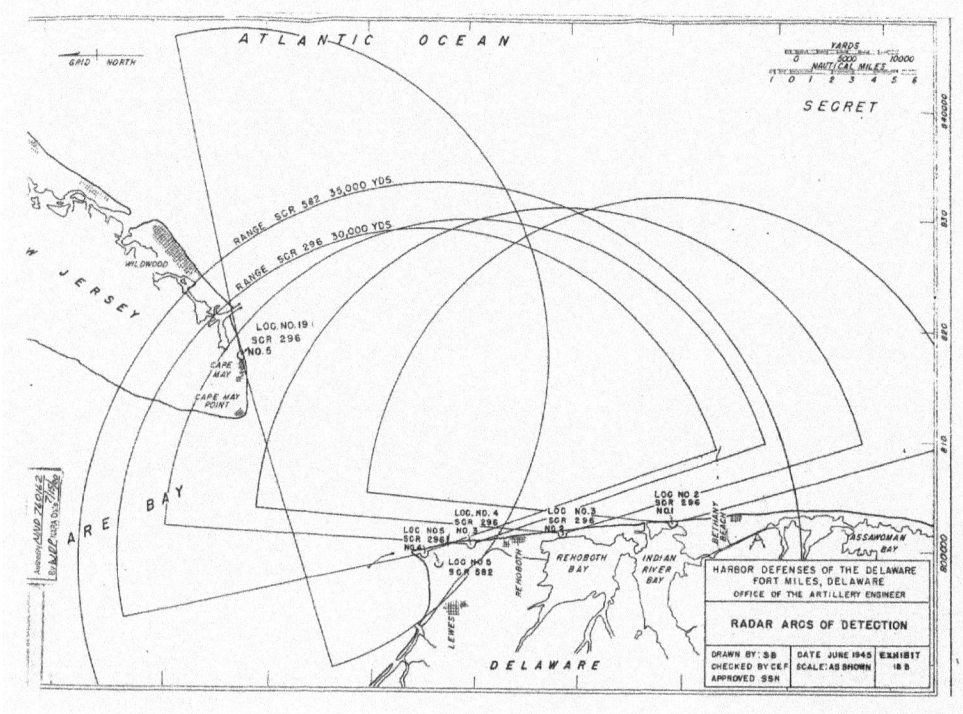

Figure 67. Radar locations and arcs of detection for the Harbor Defense of the Delaware as of June 1945. (RG177 NARA 1945)

The two fixed T3 mounts and two M1A1 mobile carriages for the M1 90 mm guns of Battery No. 7 at Cape May were emplaced on 1 June 1943 and completed on 15 June. Functional firing of Battery No. 7 was carried out on 15 October 1943, by Battery C, 21st C.A., which on 8 November 1943, began to man the battery. But, like the AMTB batteries at Fort Miles, it was not formally transferred until 21 December 1943.[124]

Many of the numerous construction projects underway since 1941 on the Delaware Capes were completed in 1943. The works at the bay's entrance were so well advanced by the end of August 1943, that HQ New York-Philadelphia Sector recommended that Forts DuPont, Mott, and Delaware be deleted from the harbor defenses. The Eastern Defense Command concurred with that recommendation on 22 September 1943 and the elimination of these forts was approved by the War Department on 23 September, effective 1 October 1943.

On that date the river forts passed to the jurisdiction of the commanding general, Second Corps Service Command.[125]

The functions associated with the tactical command and control of the harbor defenses that were carried out in the temporary HDCP were housed in Bombproof 119 (the designation for the plotting and switchboard room of BCN 519). As the casemated 12-inch battery neared completion in March 1943, preparations were made to move the command post to the permanent HDCP structure that was nearing completion. This move was made on 27 March. The new facility was of the standard splinter proof, gasproof type design used in command posts, mining casemates, etc. during World War II. It was buried in a sand dune about 4,500 feet to the rear of BCN 519.[126]

Although the HDCP was of similar dimensions to the plotting rooms of Battery Smith and Bombproof 119, its interior arrangement was somewhat different. A central corridor ran through the structure at right angles to its two

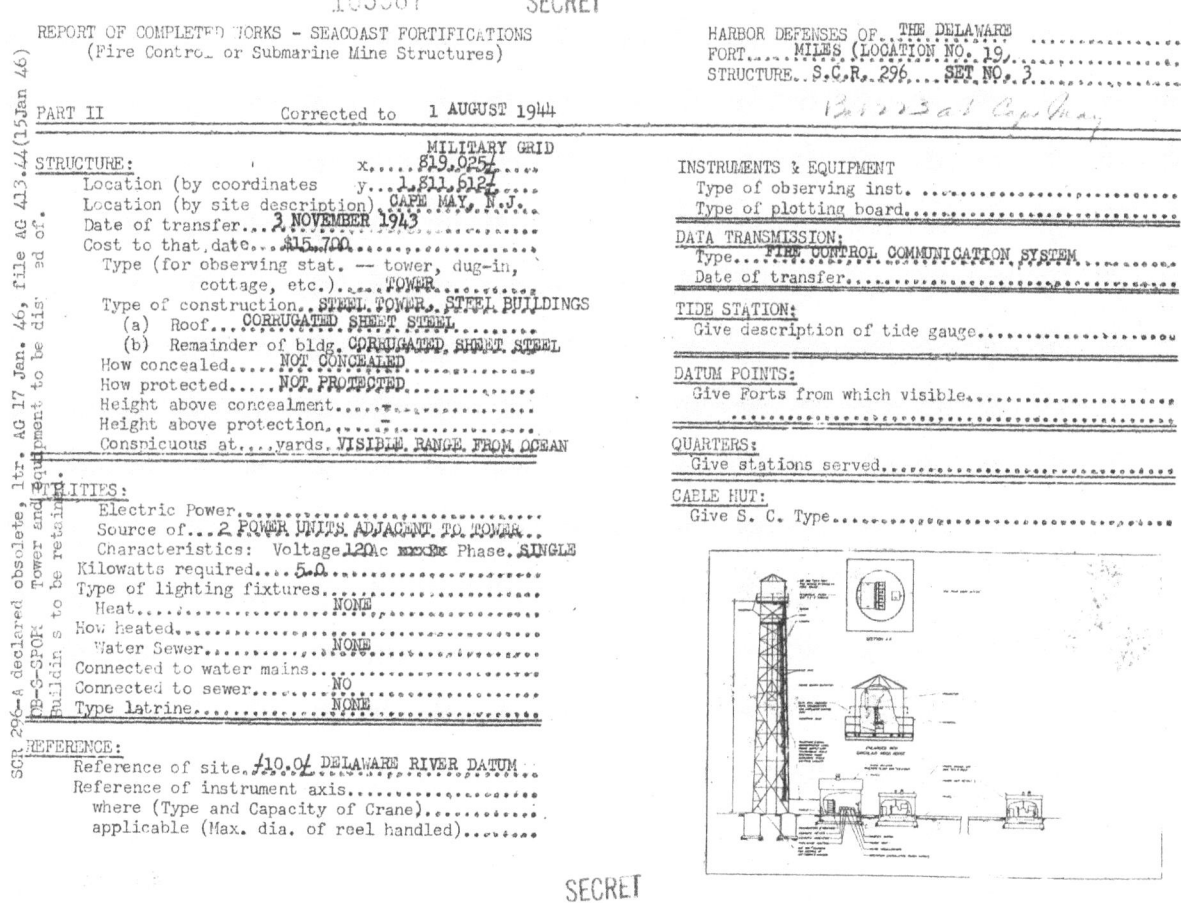

Figure 68. RCW Form 2 for SCR 296 Set No.3 (fire control) at Fort Miles, DE showing the site plan with a steel tower and radar antenna protected by cylinder structure supported by several buildings on the ground for power generation and operators. (RG77 NARA 1944)

entrances, one of which served as an emergency escape passage. The interior was partitioned into more than a half dozen rooms arranged along either side of the central corridor. These rooms provided working space for the harbor defense commander, his executive, an operations room, a message center, a room for intelligence personnel, a communications room for teletype and radio equipment, and a records room. In addition, there were rooms for the chemical warfare section (CWS) collective protectors, supplies, and equipment for gas proofing, and latrines for both officers and enlisted personnel. There was also a small emergency generator set.[127]

Near the permanent HDCP was a 95-foot, silo-type concrete fire control tower, No. 12, the harbor defense observation post (HDOP). It was also the location of the SCR-582 harbor defense surveillance radar. The radar's modulator room was located atop the concrete tower. On the roof of the modulator room, a 31-foot steel frame supported the radar antenna blister. From here the radar could clearly view the approaches and interior waters of the bay, to monitor maritime traffic.[128]

Initially the joint planning board had contemplated placing both the HDCP and the permanent HECP in a single structure near the Great Dune. By 8 May 1942 the board concluded that the Great Dune was too far from the channels of the bay for visual communication. When the Bell Haven Surf Club building was occupied as a temporary HECP in early December 1941, the board decided that the HECP should be in a tower closer to Cape Henlopen Point, for better visual surveillance of the bay's entrance and improved signal communications.[129] The HECP finally moved out of the surf club on 8 June 1943, to Fire Control Tower No. 9. This silo-like concrete tower was several hundred feet to the rear of Batteries 5, 5A, and 5B on Cape Henlopen. The tower also housed base end stations for the two 6-inch batteries under construction at

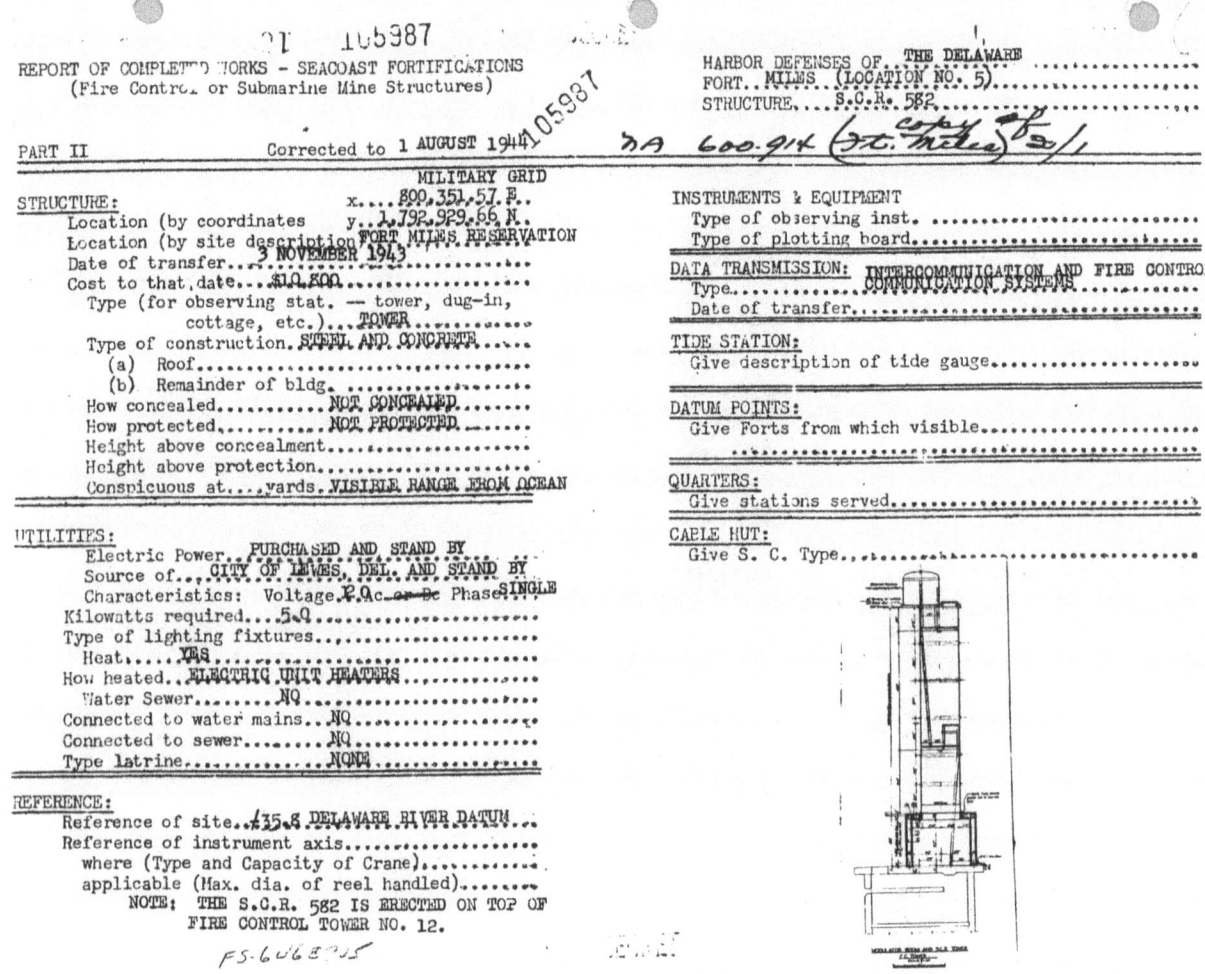

Figure 69. RCW Form 2 for SCR 582 (harbor surveillance) at Fort Miles, DE showing the site plan with a special steel and concrete tower and radar antenna located on top of Fire Control Tower #12. (RG77 NARA 1944)

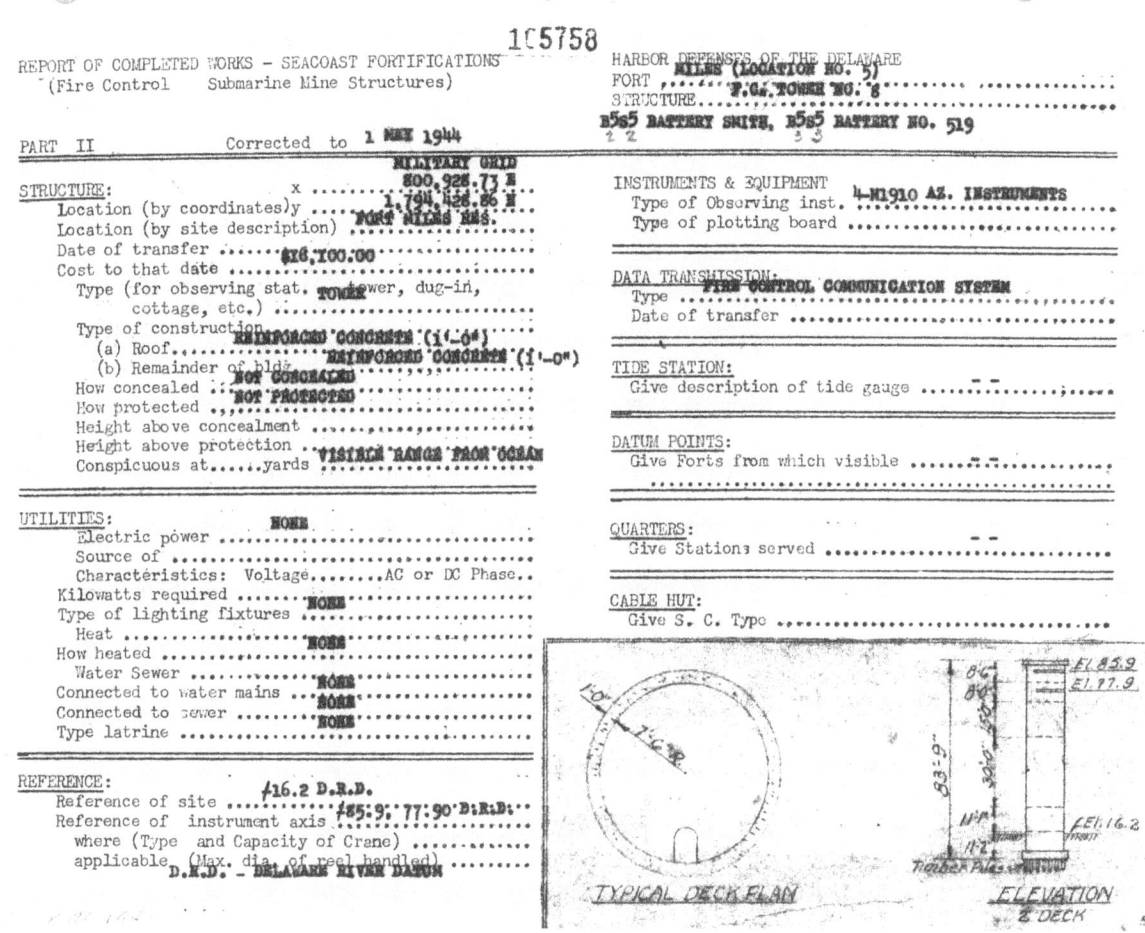

Figure 89: RCW Form 2 for Fire Control Tower No. 8 at Fort Miles, DE showing the plan for the 84-foot concrete tower with two observation levels, one assigned to Battery Smith and the other on assigned to Battery #519. (RG77 NARA 1944)

Fort Miles. The HECP itself was housed in a wooden structure on top of the tower. This served as the command post's watch room. The roof of the watch room served as the signal station platform where the signal searchlights and flag hoist bags were located. A signal mast equipped with halyards for the flag hoists and yardarm blinkers stood next to the tower.[130]

The site for the modern 6-inch battery, BCN 223, was selected by 12 January 1942, on Cape May directly to the rear of the 155 mm guns of temporary Battery 25 that it was slated to replace. Actual construction did not begin until 12 September 1942. Although the last of the modern 6-inch barbette batteries to be started in the HD of the Delaware, it was the first to be completed—23 June 1943.[131] While Batteries 221 and 222 on Cape Henlopen were provided with M1903A2 guns on M3 long range barbette carriages, Battery 223 at Cape May received the new T-2 model 6-inch guns on M3 barbette carriages. (After these new guns were approved for service use, they were designated M1 guns.) The M1 guns ranged some 26,000 yards to seaward.[132]

By the latter part of 1943, the Harbor Defenses of the Delaware had a very different appearance than that of two years before. Of the modern armament projected, most batteries were in place, and the temporary batteries were already being phased out. The advent of radar was one of the most significant developments in seacoast defense during World War II. Daylight visibility along the Atlantic seaboard rarely extended beyond ten or fifteen thousand yards. At night, effective visibility was often less than eight thousand yards even with the 800 million-candlepower 60-inch searchlights. When radar was introduced, visibility restrictions diminished, and while not wholly perfected during the war, its value became increasingly clear. On 3 November 1943 the various radar installations in the harbor defenses were placed in service. In addition to the SCR-582 harbor surveillance radar at the HDOP, provision was made for five SCR-296A fire

Figure 90: Construction at Fire Control Tower No. 3 at Dewey Beach, DE on January 23, 1942 using concrete forms that jack up the tower once the concrete layer below has hardened. (RG77 NARA 1942)

control radar sets to serve Batteries Smith, Herring, Hunter, 223, and 519. These radar sets were located along the coastline between Bethany Beach and Cape May. In addition to those two locations, one set was at Fort Miles, one was north of, and one south of, Rehoboth Beach. While each battery had one primary radar, any radar could be used by any battery.[133]

On 8 November 1943, the 155 mm batteries at Capes Henlopen and May were eliminated from the harbor defense project, their places being taken by the new 6-inch batteries. Although deleted from the project, the 155 mm guns on Cape Henlopen were not removed until 1 February and those on Cape May not until 4 February. All of these weapons were then sent to the Baldwin Locomotive Works.[134]

Calibration firing of the new 90 mm AMTB batteries was carried out between 11 and 27 November and on 21 December 1943, the three batteries were transferred to the coast artillery. On 18 December 1943, however, the New York-Philadelphia Sector of the Eastern Defense Command directed that all 90 mm guns on mobile M1A1 carriages be placed in storage. Consequently, all AMTB batteries in the harbor defenses were reduced to pairs of 90 mm guns on fixed mounts. It is somewhat ironic that just as the new AMTB batteries at Cape Henlopen were finally being placed in service to replace the 3-inch guns of Battery No. 5, the guns of Battery No. 5 and its battery commander's station, in service for a year and a half, were at last formally transferred to the garrison on 11 December 1943.[135]

Drawdown of Delaware Bay Defenses to Support Overseas Operations

In early 1944, the Harbor Defenses of the Delaware were organized as follows[136]: The HQ and HQ Battery of the Philadelphia-Delaware Sub-Sector and HD of the Delaware, 14 officers and 38 enlisted men, was at Fort Miles, along with the 240th Army Ground Forces Band, 1 warrant officer and 25 enlisted musicians. In addition to the command elements, the provisional harbor defense training and replacement battery of 273 enlisted men operated under the harbor defense headquarters.

The 21st C.A.(HD) Regiment (Type D), was composed of 29 officers, 6 warrant officers, and 621 enlisted personnel. The HQ and HQ Battery and medical detachment were posted at Fort Miles, as were Batteries A and B, who continued to operate the mine command in company with the 12th and 19th C.A.M.P. Batteries on

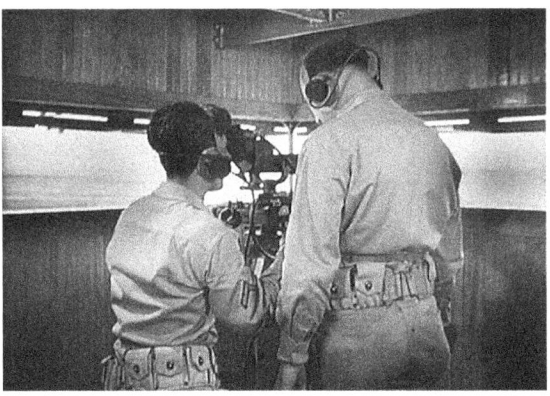

Figure 91: One of the steel fire control towers along Delaware Bay in use in 1942 with observer using an Azimuth Instrument M1910A1 for target sighting while the reader is calling off bearings via telephone to Battery Hall's plotting room. (NARA 1942)

board the mine planters Sylvester, Frank, and Casey. Battery C was at Cape May, where it manned the 155 mm guns of Battery 25 as well as the new pair of 90 mm guns of AMTB Battery 7 and was preparing to take possession of the 6-inch guns of BCN 223.

The 28 officers, 3 warrant officers, and 586 enlisted men of the 261st C.A.(HD) Bn.(Sep), were all at Fort Miles, where Battery A manned Battery Smith's 16-inch guns. Battery B was assigned to BCN 519, transferred on 13 February. Battery C manned both Battery Hunter's 6-inch guns and the four fixed 90 mm guns of Batteries 5A and 5B.

The 287th C.A.(Ry) Bn.(Sep), with its 21 officers, 3 warrant officers, and 550 enlisted men continued to man the two batteries of 8-inch railway guns at Fort Miles.

Numerous changes loomed by early 1944; some would reduce the administrative overhead in the defenses and eliminate units no longer required in the current harbor defense project. With the invasion of France imminent, there was a vital need for more troops in the Army Ground Forces. On 21 January, the War Department ordered a reduction in the Eastern Defense Command of about 60 percent and the reassignment of the excess personnel to the Army Ground Forces. G.O. No. 6, EDC, February 12, 1944, stipulated the units to be inactivated. A few days later, movement orders transferred the 287th C.A.(Ry) Bn. to Fort Jackson, S.C., to be inactivated and its personnel reassigned to the field artillery.[137] On March 2, 1944, movement orders were received from the Southeastern Sector transferring the HQ Detachment, Philadelphia Sub-Sector, along with the HQ Battery, 261st C.A. Bn., and the

Figure 92: Fire Control Tower at Big Stone Beach, DE on September 23, 1941 to support the 12-inch guns at Fort Saulsbury, DE with a concrete cable hut for receiving communication cables back to the fort. (RG77 NARA 1941)

Figure 93: Fire Control Tower at Big Stone Beach, DE on May 2, 2015 as visited by the CDSG annual conference was not as sturdy condition as it was in 1941. (McGovern Collection 2015)

Figure 94: The Fire Control Towers No. 5 and No. 6 on the beach south of Battery Herring at Fort Miles, DE have become iconic symbols of the Delaware shoreline and resort area. (DNREC 2015)

240th Army Ground Forces Band to Fort Jackson. With these departures, the firing batteries of the 261st were attached to the 21st C.A.[138]

While some of the fortification elements had been formally transferred to the coast artillery by the Corps of Engineers, one major exception was the fire control towers along the shoreline from Bethany Beach, some sixteen miles south of Fort Miles, to North Wildwood, New Jersey. These tall concrete towers had been in service for many months when they were finally formally transferred on 8 April 1944. Several days later, although they had been in service since 1942, the emplacements for the two batteries of 8-inch railway guns were also transferred. Ironically, this occurred more than a month after the positions had been vacated. Also transferred on 14 April were the now empty Panama mounts and igloo service magazines of temporary Batteries No. 22 and 25 on Capes Henlopen and May.[139]

Organizationally, after the reductions in February and March of 1944, the HD of the Delaware remained basically unchanged until October 1, 1944, when they were reorganized in accordance with G.O. No. 13, September 19, 1944. HQ and HQ Battery, 21st C.A. was redesignated HQ and HQ Battery, H.D. of the Delaware; and the inactive HQ and HQ Battery, 1st Bn., 21st C.A., was reactivated and redesignated HQ and HQ Detachment, 21st C.A. Bn. (Sep). The firing batteries of the 21st and 261st Coast Artillery were also redesignated:[140]

Figure 96: Night lighting recently installed on Fire Control Tower No. 3 at Dewey Beach, DE as shown on February 15, 2018 as part of an effort to open the tower for tourists. (DNREC 2018)

- Batteries A, B, and C, 21st C.A. Regiment, as Batt. A, B, and C, 21st, C.A. (HD) Bn.(Sep). Battery A, 261st, as Battery E, 21st C.A.(HD) Bn. (Sep), manning Battery Smith
- Battery B, 261st, as Battery D, 21st C.A.(HD) Bn. (Sep), manning Battery 519.
- Battery C, 261st, as Battery F, 21st C.A.(HD) Bn. (Sep), manning Battery Hunter
- Battery G, Seacoast Searchlight Platoon, 21st C.A. (Inactive) was activated and redesignated Seacoast Searchlight Platoon, Battery G, 21st C.A.(HD) Bn. (Sep), continuing to operate the harbor defense searchlights. The 21st C.A. Bn. manned the harbor defenses through the winter of 1944/1945. The next round of force reductions, with inevitable reorganizations, occurred in the spring of 1945. The harbor defenses were to be reduced to a single mine battery, one 6-inch battery, and one AMTB battery. All other batteries were ordered inactivated.[141]

On 1 April 1945, the HD of the Delaware received G.O. No. 1, Southeast Sector, Eastern Defense Command, March 28, 1945. The 12th C.A.M.P. Battery was to be disbanded and the mine planter *Sylvester* withdrawn from the bay, leaving only the *Frank* and its battery, the 19th, along with Battery A of the 21st to maintain the minefield.[142] The HQ and HQ Detachment and Batteries D and F, 21st C.A. Bn., were also inactivated. A second

Cape May Military Reservation, Cape May - New Jersey State Park

Battery ID	Number and type of guns	Carriage type	Service years	Notes
#223	2 x 6-inch	SBC	1944-1947	Now in surf/beach
AMTB #7	2 x 90mm	F	1943-1946	Now in surf/beach
Unnamed	4 x 155mm	PM	1942-1944	In surf
Temporary	1 x 6-inch	P	1917-1919	Coast Guard base east of town, destroyed

Fort Miles, Cape Henlopen, Mine Defenses/Mine Casemate - Delaware State Park

Battery ID	Number and type of guns	Carriage type	Service years	Notes
Smith (#118)	2 x 16-inch	CBC	1943-1948	Part of Fort Miles Museum
#119	2 x 12-inch	CBC	Not Built	Replaced by #119
#519	2 x 12-inch	CBC	1944-1948	Guns from Bty Haslet, Ft. Saulsbury – Ft Miles Museum
Unnamed	4 x 8-inch	RY	1942-1944	revetments still in place
Unnamed	4 x 8-inch	RY	1942-1944	Covered/buried
Temporary	1 x 6-inch	P	1917-1918	Buried - Gun from Ft. St. Phillip
Herring (#221)	2 x 6-inch	SBC	1944-1948	Earthen cover removed, SOSUS Terminal
Hunter (#222)	2 x 6-inch	SBC	1943-1947	Navy Radio Station/Mine Control
Exam	4 x 3-inch	P	1942-1946	Mostly buried
AMTB #5A	2 x 90mm	F	1943-1946	Mostly buried
AMTB #5B	2 x 90mm	F	1943-1946	1 block covered by parking lot
Unnamed	4 x 155mm	PM	1942-1944	Two PM visible

Table 3 Summary of coastal defences and current status (SBC – Shielded Barbette Carriage; F – Fixed Mount; PM – Panama Mount; P – Pedestal Mount; CBC – Casemated Barbette Carriage; RY – Railway Carriage)

G.O. No. 3, Southeast Sector, 29 March 1945, redesignated Batteries A, C, and E, and the searchlight platoon, 21st C.A. Bn., as batteries of the HD of the Delaware, with the same designations. In addition, the HQ and HQ Battery, HD of the Delaware, was reorganized.[143] These reorganizations resulted in the following manning:

– Battery A, HD of the Delaware, stationed at Fort Miles, was assigned as the HD mine battery, with 5 officers, 1 warrant officer, and 155 enlisted men.
– Battery C, HD of the Delaware, also stationed at Fort Miles, manned AMTB Battery 5A's two 90 mm guns with 3 officers and 127 men. AMTB Battery 5B's 90 mm guns were the examination battery, also manned by a detachment of Battery C.
– Battery E, HD of the Delaware, 4 officers and 101 enlisted men at Fort Miles, manned the 6-inch guns of Battery Hunter. In addition, it provided caretaking detachments for Batteries Smith, Herring, and 519, and for Battery 223 at Cape May.
– The seacoast searchlight platoon of Battery G, HD of the Delaware, 1 officer and 32 men, manned the remaining handful of 60-inch seacoast searchlights.

This organizational structure was retained through the end of the war in August 1945. On 18 August 1945, the HDCP at Fort Miles was closed.[144] By this time, coast artillery troops at Fort Miles had dwindled to only a few hundred. There were however, other troops posted at Cape Henlopen. During the latter months of the war, the post had become headquarters for the several prisoner of war camps located in the area. Many of the Italian and German POWs worked on various parts of the reservation. In the initial postwar months, as the prisoners were repatriated, the pace of activity at the post continued to decline. By 1946, the fort's primary function had become the storage of ammunition.

The coast artillery presence at Fort Miles was in its last

The Coastal Defenses of the Delaware Bay during World War Two

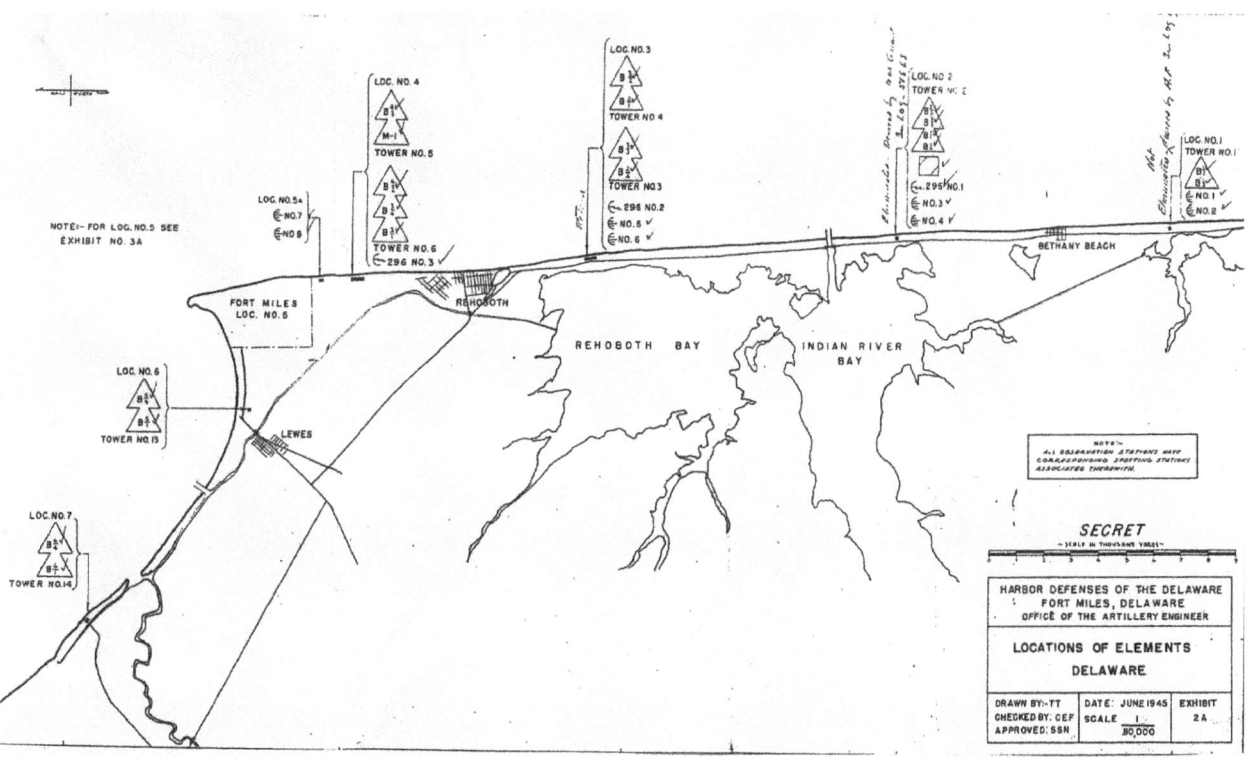

Figure 97: Location of Elements of the Harbor Defenses of the Delaware in June 1945 showing the location of fire control towers along the Delaware shoreline (not including the tower within Fort Miles reservation). (RG177 NARA 1945)

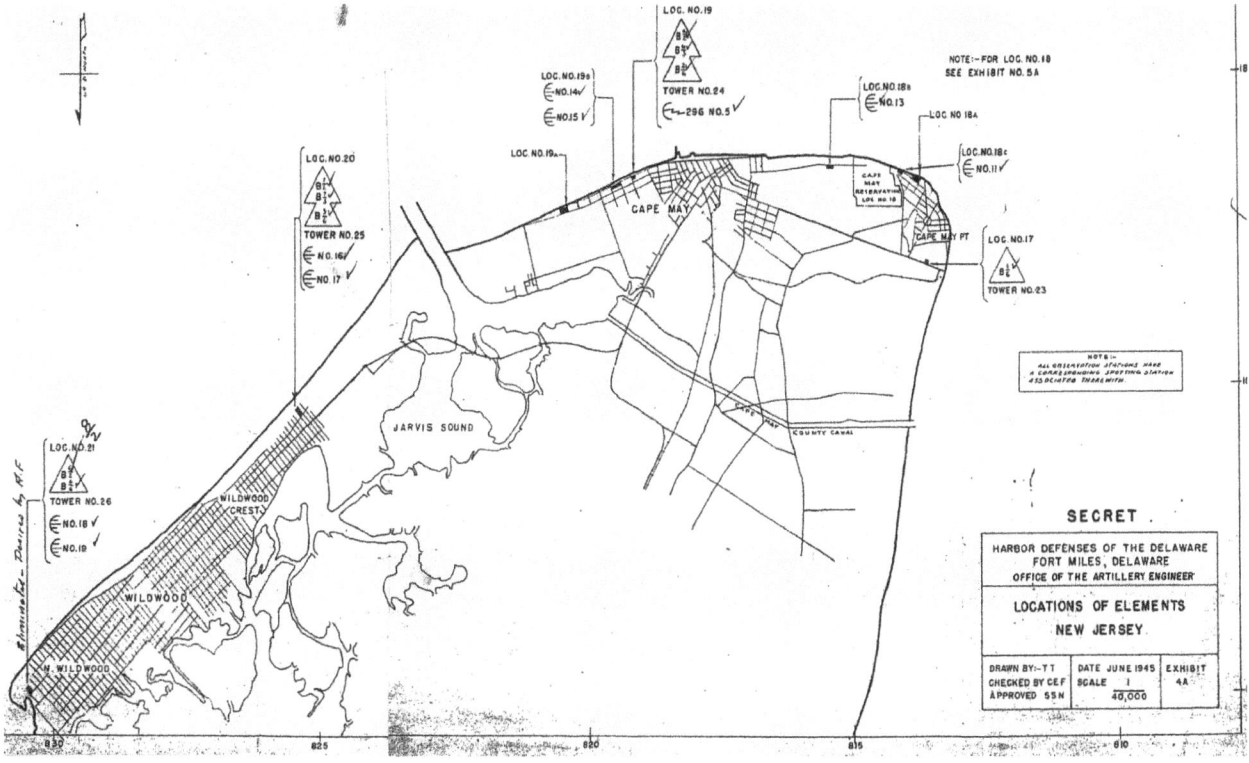

Figure 98: Location of Elements of the Harbor Defenses of the Delaware in June 1945 showing the location of fire control towers along the New Jersey shoreline. (RG177 NARA 1945)

Figure 95: The newly installed spiral staircase in Fire Control Tower No. 23 at Cape May, NJ to allow tourist to climb to the observation stations at the top the tower. (McGovern Collection 2013)

days by early summer, 1946. On March 1, 1946, the HD of the Delaware were relieved from assignment to the Eastern Defense Command and reassigned to First Army. On March 15th, the elements of the HD of the Delaware were placed under the 39th HQ and HQ Detachment, Special Troops, First Army. On 30 June 1946, HQ and HQ Battery, and Batteries A, C, and E, and the searchlight platoon of Battery G, HD of the Delaware, were inactivated and their personnel reassigned to coast artillery caretaking detachments within First Army.[145] Summarizing the coast defenses of Fort Miles and the Cape May Military Reservation during World War Two and their current status is listed below:[146]

Delaware Bay Defenses after World War Two

At the end of the war, the War Department policy was to maintain the seacoast armament, reduced, however, to standby status, while the submarine mine fields had been cleared. Subsequently, The War Department Seacoast Defense Armament Board, formed in July 1946 and presided over by Brig. Gen. Rollin L. Tilton, appraised all harbor defenses, including Delaware Bay. The board was directed to examine all seacoast armament and determine those that could be eliminated and what obsolete equipment was to be removed and what was to be stored. In its report that itemized the status of the various elements of the defenses, the board recommended that the modern armament of Battery Smith, Battery 519, and Batteries Herring and Hunter at Fort Miles, and 223 at Cape May be retained, along with the 90 mm AMTB batteries. It also noted that eleven M4 mines still remained to be raised from the minefield. The fire control instruments had all been placed in storage and many of the fire control stations built during the war were, in the view of the board, no longer required in the HD project, as they had been displaced by the five SCR-296A fire control radars. The SCR- 682 harbor defense radar at Tower 12, although still in place, needed repairs and unmanned because of a lack of qualified operators. The Tilton Board also reported that while Fort Miles was very adequately provided with barracks and officers' quarters, it was lacking in family quarters, and noted that married quarters would be required if the post was to be provided with a garrison.[147] The recommendations of the Tilton Board were not accepted by the War Department and most Army defenses were placed in caretaker status. In 1948, the Army decided to discontinue its coast artillery role for the entire country and by 1950 the Coast Artillery Corp itself was dissolved. At Fort Miles, the batteries had their armaments scrapped and the Army's remaining harbor defense responsibilities were transferred to the Navy in 1949.

While the Army used Fort Miles as a training site and for ammunition storage leading up to the outbreak of the Korean War, the Navy used the fort to support their expanded harbor defense role, as well as to establish testing stations for the early development of missiles, such as Operation Bumblebee at Herring Point. As the Army began to place less emphasis on Fort Miles after World War II, the Navy began to re-establish its presence. In 1949, the Army turned over their responsibility for controlled mine field operations to the Navy which the Navy combined with their existing harbor defense roles. The Navy used the former Army controlled mine complex,

Figure 100: The Navy installed two large troposcatter microwave antennas, each about 120 feet tall, at Battery #519 in 1964 as the Navy was using the former Army battery as the Naval Radio Station Lewes (in the foreground is parking lot for the newly formed Cape Henlopen State Park). (DNREC 1967)

which included the mine wharf, mine storehouse, and mine vessels. The Navy did not use the former Army mine casemate (this structure was transferred to State of Delaware in 1958 as a civil defense facility). Instead, the Navy used the former 6-inch Battery Hunter as their mine command post and plotting room. The Navy also used the former 16-inch Battery Smith to store spare mines in its northern gun casemate. [148 149 150]

During the 1950s, the Army was still utilizing Fort Miles as an Army anti-aircraft training facility and as overflow housing of soldiers being released from active duty after the Korean War. Its primary Army function was a morale, welfare and recreation (MWR) area for active and retired military personnel and their families. Some of the temporary hospital buildings and quarters on the post were converted to married and family housing. This function was known as the First Army Recreation Area (roughly 190 acres), a sub-post of Fort Meade, located in Maryland. Following the Korean War, a small cadre of a few hundred regulars continued to operate Fort Miles. In the late 1950s, some of the unused land and facilities were made available to the State of Delaware, as the Army decided to close Fort Miles in 1958. The closure of Fort Miles was projected to save the Department of the Army approximately $300,000 per year.[151 152]

Although the Army transferred most of its military personnel to other bases, they retained some land for defense purposes, such as Gap Filler Radar Station (1959 to 1963) located at first on top of FC Tower #12 and then relocated to the former Battery 519. In 1956, the Army transferred to the Navy parcels of land containing 593.46 acres, which would be used for several naval functions over the next 30 years. The Army would also convey about

Figure 99: Testing of experimental anti-aircraft rockets at Fort Miles, DE as part of Operation Bumblebee by the Navy to develop surface-to-air missile defenses for warship in 1946. (DNREC 1946)

554 acres to the State of Delaware, which would become the future Cape Henlopen State Park. On 17 October 1964 the State formally accepted the former Fort Miles property as Cape Henlopen State Park. To reduce maintenance costs, many of the buildings were either sold and moved or torn down.[153 154 155]

The Navy continued to research and develop technology to identify and expose the location of ships and, in particular, submarines during this period. The system they established to detect submarines was known as Sound Surveillance System (SOSUS). SOSUS systems consisted of bottom-mounted hydrophone arrays connected by underwater cables to facilities ashore. This enabled the detection of submarines by their faint acoustic signals. In 1961, on the southern end of the former Fort Miles reservation the Navy constructed a SOSUS Naval Facility (NavFac Lewes), which was being relocated from Cape May, New Jersey due to storm damage to a similar facility located there. The facility was completed in 1962. Additionally, that year, a Naval Radio Station at former Battery Hunter was built-out and completed in 1963 (later moved in 1964 to Battery 519).[156 157] A microwave antenna was placed on top of the bunker. This was to be a tropospheric scatter, or "troposcatter" radio station. The bunker of Battery Hunter was only an interim site for the station, because of the need for higher power transmissions, more spacious equipment areas and larger antennae were required. In 1964, the station was moved to the former 12-inch Battery 519. By 1970, satellite communications capabilities had negated the requirement for troposcatter communications. The radio station and the ships were decommissioned. The antennae were removed from Cape Henlopen soon thereafter.[158 159]

In March 1981, the Department of Defense (DoD) announced that the SOSUS Naval Facility would be closing. The SOSUS Naval Facility Lewes closed on 30 September, 1981 and the Naval Reserve Training Facility (NAVRESFAC) Lewes was established shortly after under a compromise with the State of Delaware to allow the Naval Reserve to remain at the Lewes site. As the largest

Figure 101: Cape Henlopen State Park is Delaware's most popular state park due to its beautiful beaches so to accommodate those visitors DNREC has built facilities like the beach house in this aerial view that straddles the former Battery #22 (155mm GPF) at Fort Miles, DE with the battery's Panama Mounts still in place. (Williford Collection 2012)

Figure 102: In 2004, the Fort Miles Historical Association (FMHA) undertook a project to turn the abandoned Battery #519 at Fort Miles, DE into military museum with the goal to restore one of 12-inch gun casemates to its wartime appearance by placing a 12-inch gun back into position and as this August 22, 2009 photograph shows the great results. (McGovern Collection 2009)

reserve unit with a full-time presence and an increasing amount of equipment, the MIUW took over the former NavFac Auxiliary Building (former Battery Smith) as its base. The former NavFac Multi-purpose building (Biden Center) became the headquarters of the NAVRESFAC and several other reserve units began to train there. The Navy transferred the rest of its holdings to the State of Delaware by 1983, leaving a balance of 16.8 acres at the southern end.[160] [161] However, in 1995, it was announced that Senator Biden was seeking the return of the land on which the Naval Reserve Facility was located for inclusion in Cape Henlopen State Park. Despite the rapid rise in activity of Naval Reserve Facility Lewes, the Navy could no longer justify keeping the facility open. So, in 1996, after almost 100 years at Cape Henlopen, the final military units left and the Navy transferred the remaining land to the State of Delaware.

Delaware Bay's World War Two Defenses Today

Today, the Delaware Department of Natural Resources and Environmental Control (DNREC) operates and maintains most of the remaining former Fort Miles buildings at Cape Henlopen State Park. Additionally, the Fort Miles Historical Association, formed in 2003, works in partnership with Delaware State Parks to assist with the development and operations of a military museum and historic area. In 2004, the Fort Miles Historic Association (FMHA) established the Fort Miles Museum within the former Battery 519 (two 12-inch guns) which features exhibits (including a 12-inch gun mounted in the gun casemate) and an outdoor Historic Area with large artifacts related to the wartime history of the Fort Miles and the Delaware Bay.[162]

For hundreds of years, Cape Henlopen has moved

Figure 104: The carriage of the 12-inch gun was formerly use a testing girder and slide for naval guns at the Dahlgren NSWC in Virginia, while the 12-inch/50 Mk7 barrel used to be mounted on the USS Wyoming (B-32). (McGovern Collection 2019)

Figure 103: The view from within Battery #519's gun casemate #1 at Fort Miles, DE on May 19, 2016 with a glass wall to control humidity and sand from coming into the new museum. (McGovern Collection 2016)

Figure 105: Battery #519's service corridor at Fort Miles, DE restored to its World War Two appearance with a replacement overhead rail system to move 12-inch projectiles from the shell room to the breech of the 12-inch gun. (McGovern Collection 2019)

Figure 106: The FMHA has a volunteer group called the "Bunker Busters" that is responsible for much of work in restoring Battery #519 at Fort Miles, DE and they have crafted replicas seen in this image - shot cart, overhead rail system, shell hoist, and even the wooden 12-inch shells (easier for demonstrating use). (McGovern Collection 2016)

Figure 107: The rooms in Battery #519 at Fort Miles, DE have been transformed by the "Bunker Busters" into displays on various coast artillery artifacts, such as this view of controlled mines used at the beginning of WW2 in Delaware Bay. (McGovern Collection 2020)

Figure 108: The Fort Miles Museum in Battery #519 at Fort Miles, DE includes an art gallery for paintings that recorded the military activities in Delaware Bay area during WW2. (McGovern Collection 2016)

Figure 111: As part of Fort Miles Museum programing the park's staff provides demonstrations of a gun drill on a naval 3-inch gun including firing (no shell!) several times a week. (DNREC 2018)

Figure 109: One of the largest projects undertaken by the FMHA was the movement and display of 16-inch/50 Mk 7 barrel (similar to the 16-inch guns that were once installed in Battery Smith) mounted on a proof girder and slide from Dahlgren NSWC which took place on May 20, 2016 using a 500-ton hydraulic crane seen in this image. (McGovern Collection 2016)

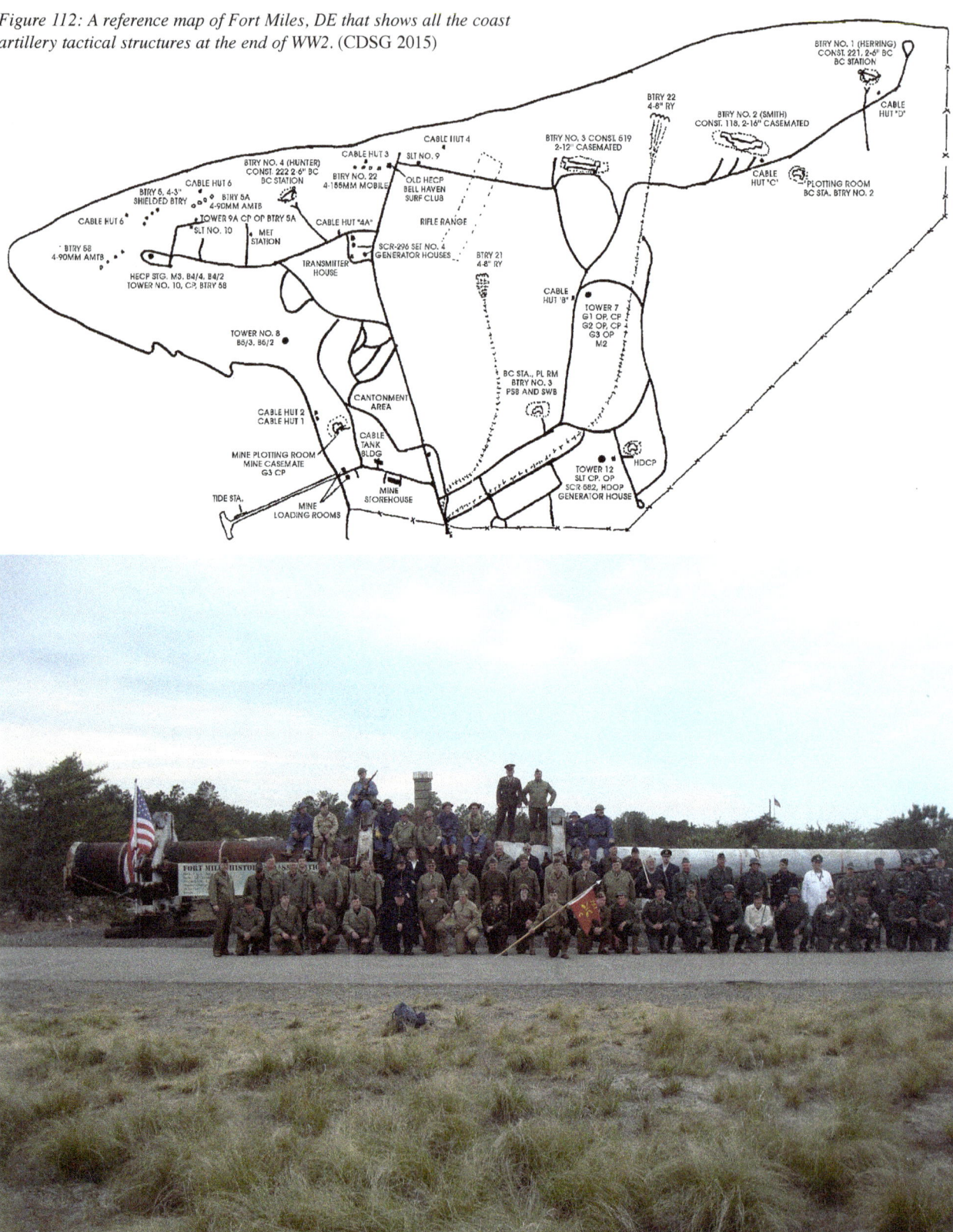

Figure 112: A reference map of Fort Miles, DE that shows all the coast artillery tactical structures at the end of WW2. (CDSG 2015)

Figure 110: The FMHA and DNREC arrange a range of historical events and tours during the tourist season at Cape Henlopen State Park as seen in this image were a wide range (not sure how British policemen fit into Fort Miles' history) of reenactors gathered for a group photo by the 16-inch/50 barrel (before it was mounted). (McGovern Collection 2012)

westward at an average rate of four to five feet per year. Consequently, the configuration of today's Cape Henlopen has substantially changed from the early 1940s. For example, when construction of the temporary platforms for the pair of 12-inch guns was begun in 1941 on Cape Henlopen, they were some 200 feet from the water's edge. Today they lie in the surf, visible only at low tide. In spite of the erosion of the shoreline, many of Fort Miles' fortifications still exist. Among the more obvious are the modern primary and secondary gun batteries of the 1940s Modernization Program. While the interiors of most of these were modified considerably by the Navy after World War Two, their basic design can still be discerned. Battery Herring has lost its earthen cover when converted to a SOSUS terminal facility. While the plotting spotting room of Battery Smith is still extant, its interior has been modified and it is not open to the casual visitor. Both Battery 519 and Battery Smith are used by the FMHA as a museum and workshop respectively. Battery 519's plotting spotting room is visible from the main access road but is sealed to prevent access by vandals. Battery Hunter remains and its battery commander's station is used as seabird observation platform. One of the gun blocks for AMTB Battery 5A is still visible at Cape Henlopen Point, although those for the 3-inch guns and AMTB Battery 5B appear to be buried beneath the shifting sands of the cape.

The old Bell Haven Surf Club was dismantled long ago, but two of the semicircular Panama mounts of Battery No. 22 can still be found in the vicinity of the park's modern bathhouse. The cement bag epaulements for the 8-inch railway guns can still be seen in the wooded area south of the old parade ground and on the north side of the Great Dune. The foundations of the old mine structures and numerous other buildings can be found in the thickets of beach heather and other vegetation, while the Mine Storehouse is used today by the University of Delaware while the Mine Casemate is sealed to protect it from vandals. Nestled along a little-traveled trail that once paralleled a branch of the post railroad are a series of large storage magazines that stored reserve ammunition for seacoast batteries and the controlled mines.

Perhaps the best recognized remnants of Fort Miles are

Figure 49: Aerial view of Fort Saulsbury at Slaughter Beach, DE on November 14, 2012 with Batteries Hall and Haslet being used for storage. (Williford Collection 2012)

the ubiquitous silo-type fire control towers that abound along the Delaware and New Jersey shore. Some sixteen of these circular concrete towers were erected in the early 1940s, between Bethany Beach, Del., and North Wildwood, N.J.; seven of them on or adjacent to Fort Miles. FC Tower No. 7, near the Great Dune, has been modified to have a steel spiral staircase to allow visitors to climb to the top of the tower. FC Tower No. 9, used to support the HECP, was taken over by the Delaware Pilots Association, which used the old wooden structure until 1986, when it was replaced by a modern concrete pilot station atop the tower. This tower is not open to the casual visitor. Efforts are underway to restore FC Tower No. 3 which is located within the Delaware Seashore State Park on Route One just south of Dewey Beach, DE to allow visitor to access the tower in a similar way as FC Tower No.7.

Cape May too, is drifting northward. BCN 223, once some 550 feet from the surf line, has varied from being awash in the surf allowing visitors to walk under the emplacement (which is supported by piles) to sitting firmly on beach with sand back up to its doorways. It too was modified after World War II as part of the navy's undersea monitoring (SOSUS) system, until this function was transferred to Fort Miles. FC Tower No. 23 in Cape May was modified in 2009 to add a spiral staircase, lighting and safety features allowing the public to climb to the top of the tower and a wooden walkway was built to Sunset Avenue to access the tower. Farther up Delaware Bay, more evidence remains of the bay's military heritage. At Fort Saulsbury, the two batteries for the long-range 12-inch guns have remained intact and basically unaltered structurally, although without their armament, is in private hands and not open to the public. One of the 1940's steel

Figure 50: Western entrance to Battery Hall at Fort Saulsbury, DE during the CDSG annal conference visit on May 2, 2015. (McGovern Collection 2015)

Figure 51: Interior of shell rooms and powder magazines of Battery Hall at Fort Saulsbury, DE during CDSG annual conference on May 2, 2015. (McGovern Collection 2015)

Figure 52: Shaft for spiral staircase (missing but handrail still in place) that allows access from one of Battery Hall's plotting room to its battery commander's station at Fort Saulsbury, DE. (McGovern Collection 2015)

Figure 53: CDSG members explore Battery Hall's mechanical indicator display and tunnel that transmitted range and azimuth from the plotting room to firing emplacement at Fort Saulsbury, DE on May 2, 2015. (McGovern Collection 2015)

Figure 54: Aerial view of Battery Hall at Fort Saulsbury, DE on November 14, 2012 with its two 12-inch gun emplacements while its two battery commander stations (and connecting trench) are located on the roof of it protected structure. (Williford Collection 2012)

Figure 56: Aerial view of Battery Smith at Fort Miles, DE on November 14, 2012 with its two 16-inch gun casemates with its protected service corridor between them (the 1962 Navy barrack building is seen in the background). (Williford Collection 2012)

Figure 57: Inside the gun casemate #2 for Battery Smith at Fort Miles, DE on May 3, 2015 during the CDSG annual conference (now used as workshop by the Fort Miles Historical Assn.). (McGovern Collection 2015)

FC Towers remains in weathered condition at Big Stone Beach. Further up the Bay, the emplacements for 3-inch rapid fire guns at the Liston Range Light remain, as well as the three forts guarding the Delaware River (Fort DuPont, Fort Delaware, Fort Mott).

Conclusion

The investment in the construction and manning the coastal defenses of the Delaware Capes during World War Two turned out not to be needed, yet when designed in 1940-41 the threat to the Eastern Seaboard appeared to be a real possibility. France had fallen to Germany and the United Kingdom was under threat of invasion. If the French and English navies became controlled by Germany, the prospect of naval attacks of America's Atlantic coast became much more imaginable and the urgency to emplace large caliber coast artillery and minefields to fend off these attacks becomes a reasonable

Figure 58: The outside of gun casemate #1 for Battery Smith at Fort Miles, DE on May 3, 2015 (the rolling door and cement blocks enclosing the gun block were add by the Navy during the 1960's to allow the space to serve as a vehicle repair facility) as a CDSG member checks out several Naval barrels. (McGovern Collection 2015)

Figure 59: Main service corridor of Battery Smith at Fort Miles, DE on May 3, 2015 as CDSG members explore the shell rooms and powder magazines along the corridor (later use by the Navy resulted in the blue paint scheme). (McGovern Collection 2015)

course of action. As it turned out the German Navy conducted submarine warfare within the sight of these defenses. The existence of controlled mines, rapid-fire guns, detection devices, aircraft patrols, and naval vessels deterred the U-Boats from entering the Bay (along with the dangers of shallow waters of the Bay for a submarine). Once the US Army and Navy assessed the threat of attack to be low, they drew down the assigned service personal and replaced those remaining with limited-service personnel. Once World War Two ended, the lack of any naval threat to the American homeland allowed for the complete abandonment of coastal defenses even though some were only completed and transferred a few weeks prior to their abandonment.

Public interest in America's historic coast defenses has grown substantially during the last 50 years, as former military posts became parks with public access. The publication of *Seacoast Fortifications of the United States:*

Figure 65: Aerial view of Battery #519 at Fort Miles, DE on November 14, 2012 with its two 12-inch gun casemates with connecting service corridor while in the background the battery's cantonment area and Fire Control Tower #7. (Williford Collection 2012)

Figure 66: The uncovered (after backhoe removed earth fill) entrance to Battery #519 Plotting and Switchboard (PSR) Room at Fort Miles, DE during the CDSG annual conference on May 2, 2019. (McGovern Collection 2015)

Figure 67: Inside the Plotting and Switchboard Room (PSR) for Battery #519 at Fort Miles, DE on May 2, 2015 visited for the first time in years by the CDSG annual conference. (McGovern Collection 2015)

Figure 77: Aerial view of Battery #223 at Cape May Military Reservation, NJ on November 14, 2012 showing how the shoreline has moved resulting in the 90mm and 155mm emplacements disappearing into the sea with only the protected emplacement of Battery #223 surviving

Figure 80: CDSG members explore the service corridor of Battery Hunter (#222) on May 2, 2015, the rooms off the corridor were for the storage of 6-inch shells and powder canisters. (McGovern Collection 2015)

Figure 81: Aerial view of Battery Hunter (#222) at Fort Miles, DE on November 14, 2012 showing the bird watching platform built on top of the battery's battery commander's station. (Williford Collection 2012)

Figure 84: Battery Herring at Fort Miles, DE was used by the Navy from the 1960's to 1990's to support a SOSUS station (using hydrophone to listen for Soviet submarines) so they removed the battery's earthen cover and created additional access by cutting through cement and rebar as seen in this image. (McGovern Collection 2015)

An Introductory History, by E.R. Lewis in 1970 was a pivotal event, giving the first well-documented interpretive history of American coast defenses; it remains the basic study in the field. The Coast Defense Study Group (www.CDSG.org) was organized in 1985, and their annual conferences, *Coast Defense Journal*, *CDSG Newsletter*, website, and reprints of key coast defense books have fostered interest in American coast defense and assisted both the public and park staffs to understand these

Figure 83: A rare image of the 6-inch M1903A2 gun and shielded barbette carriage M1 of Battery Herring (#221) at Fort Miles, DE. (DNREC 1944)

defenses and how to interpret their surviving elements. It is hoped that this interest will translate into efforts to preserve and restore these sites for current and future generations.

Biographical Note

Terrance McGovern has authored eight books and numerous articles on fortifications, four of those books being for Osprey's Fortress Series (*American Defenses of Corregidor and Manila Bay 1898- 1945*; *Defenses of Pearl Harbor and Oahu 1907-50*; *American Coastal Defenses 1885-1950*; *Defenses of Bermuda 1612-1995*). He has also published 12 books on coast defense and fortifications through Redoubt Press or CDSG Press. Terry was Chairman of the US-based Coast Defense Study Group and continues to be a long-time officer. He has also been the Editor of the Fortress Study Group annual journal, *FORT*. He is a director of the International Fortress Council and the Council on America's Military Past. He can be contacted at tcmcgovern@att.net

Bibliography

Primary Sources
PUBLISHED MATERIAL

Conn, S. Engelman, R.E. and Fairchild, B. (1964) *Guarding the United States and Its Outposts*. Washington, Department of the Army.

Bronk,W.M. 1st Lt. (1945) *A History of the Eastern Defense Command*. New York

Coast Defense Study Group, (2015) *Conference Notes for the 2015 Annual CDSG Conference to the Harbor Defense of Delaware Bay*. McLean VA, CDSG.

Gaines, W.C. (1996) 'The Maritime Defense of the Delaware, 1771-1950" Coast Defense Journal. CDSG.

Manthorpe, W.H.J. Jr. (2014) *A Century of Service: The US Navy on Cape Henlopen, Lewes, Delaware: 1898-1996*. Wilmington, DE, Cedar Tree Books, Ltd.

US Army (1945) *Harbor Defense Project Delaware Bay, Annex A, Supplement to the Harbor Defense Project, Harbor Defenses of Delaware Bay*. Fort DuPont, DE.

Warrington, C. W. (2003) *Delaware's Coastal Defenses:*

Figure 85: Aerial view of Battery Herring (#221) at Fort Miles, DE on November 14, 2012 showing its gun emplacements and battery structure without its earthen protection (and BC station is missing) as the Navy built a SOSUS listening station in front of the battery (metal building was torn down but its foundation remains). (Williford Collection 2012)

Fort Saulsbury & a Mighty Fort Called Miles. Delaware Heritage Press.

Wray, G. and Lee Jennings, L .(2005) *Fort Miles*. Charleston, SC, Arcadia Publishing Co. Inc.

Zink, R. (1996) '*The Forts of "Wherever" – Defending the Delaware River*. Coast Defense Study Group News.

ARCHIVAL MATERIAL

Harbor Defenses of the Delaware: *Fort Miles, Delaware*, RG 77, Box 1-65, Entry 90, NARA Branch Depository, Philadelphia, PA.

United States Department of the Interior, National Park Service, NATIONAL REGISTER OF HISTORIC PLACES REGISTRATION FORM - *Fort Miles Historic District CRS# S-6048*

Ross, E.G.R *FORT MILES: ADAPTATION, RESILIENCE AND A MILITARY LEGACY* A thesis submitted to the Faculty of the Univ. of Delaware in partial fulfillment of the requirements for the degree of Master of Arts in Urban Affairs & Public Policy Spring 2002

Notes

1. WorldAtlas, Delaware Bay, St. Laurent, Quebec, 2020, webpage (https://www.worldatlas.com/articles/delaware-bay-united-states.html)
2. The Encyclopedia of Greater Philadelphia, *"Forts and Fortifications by Jeffery M. Dorwar"*, Rutgers, The State University of New Jersey, Camden, NJ, webpage (https://philadelphiaencyclopedia.org/archive/forts-and-fortifications/)
3. Coast Defense Study Group, "*Forts, Military Reservations and Batteries 1794-1945*", CDSG website 2020, webpage (https://cdsg.org/fort-and-battery-list/)
4. ARCE, 1872, p.13. RCW-Del, Beach Avenue, near Cold Spring Inlet, Cape May, NJ; 1 6-inch B.L. Rifle, Corrected to June 1919. RCW-Del, Cape Henlopen, DE, 1 6-inch B.L. Rifle, Corrected to June 1919. Order of Battle, Vol III, Pt. 2, p.1151.
5. RCW-Del, Location No. 18, Engineer Depot, Wilmington, 2 3-inch Anti-aircraft Gun Emplacements, Corrected to June 1919. RCW-Del, Location No. 13, Marcus Hook, PA, 2 3-inch Anti-aircraft Gun Emplacements, Corrected to June 1919. RCW-Del, Location No. 15, Pretty Island, Philadelphia, PA, 2 3-inch Anti-aircraft Gun Emplacements, Corrected to June 1919. RCW-Del, Location No. 16, Oppo- site Schuykill Arsenal, Phila. PA., 2 3-inch Anti-aircraft Gun Emplacements, Corrected to June 1919.
6. RCW-Del, Location No. 6, Thompson's Point, NJ, 2 3-inch Anti-aircraft Gun Emplacements, Corrected to June 1919. RCW-Del, Location No. 5, Carney's Point, N.J., 2 3-inch Anti-aircraft Gun Emplacements, Corrected to June 1919. RCW-Del, Location No. 8, Carney's Point, NJ, 2 3-inch Anti-aircraft Gun Emplacements, Corrected to June 1919. RCW-Del, Location No. 21, Hog Island, PA, 2 3-inch Anti-aircraft Gun Emplacements, Corrected to June 1919.
7. RCW-Del, Fort DuPont, Anti-aircraft Mounts, Corrected to September 30, 1924.
8. C.W. Warrington C.W (1992) *Fort Saulsbury* pp.1-8.
9. RCW-Del, Fort Saulsbury, Battery Haslet, Corrected to November 10, 1921. RCW-Del, Fort Saulsbury, Battery Hall, Corrected to November 10, 1921.
10. 1st Lt. William M Bronk, *A History of the Eastern Defense Command* (New York, 1945), (hereafter *History, EDC)*, p.180.
11. RCW-Del, Fort Saulsbury, Battery Hall, Corrected to November 10, 1921.
12. RCW-Del, Fort Saulsbury, Battery Hall, Corrected to November 10, 1921.
13. Warrington, *Fort Saulsbury*, p.22.
14. Ibid. p.18.
15. Basic Harbor Defense Project, 1933, Harbor Defense Projects for Harbor Defenses Included in the Philadelphia-Delaware River Area, Records of the Chief of Coast Artillery 1878-1942, RG 177, National Archives, Washington, D.C. (hereafter 1933 Project)
16. Lewis Cass to Martin Van Buren, President of the Senate, Report on the Erection of a Mole and Fortification for the Protection of the Delaware Breakwater, January 26, 1836, *ASPMA*, Vol. VI, pp.22-23.
17. 1933 Project. *EDC*, pp. 145-146. *History of the Eastern Defense Command and the Defense of the Atlantic Coast of the United States in the Second World War* (hereafter *EDC*), pp.182-185.
18. *EDC* p.146. 1933 Project.
19. 1933 Project, p.12.
20. Ibid, p.15.
21. Sawicki, *AA Bns*, Vol. I, pp.430-431.
22. 8-inch Railway Guns, "52nd Coast Artillery at Lewes," New York Times, August 2, 1941, p.7.
23. Historical Data Sheet and Station List (hereafter Data Sheet), Battery E, 7th C.A.(HD) Regt., HD of the Delaware.
24. Data Sheets, 21st C.A.(HD) Regt. and 7th C.A.(HD) Regt. *History of New York-Philadelphia Sector* (hereafter *NY-Phil*), p. 79.
25. *NY-Phil*, p. 79. "Coast Artillery Orders," *CAJ*, Vol. 83, No. 5 (September-October 1940), p. 487, 488. Data Sheet, 21st C.A.(HD) Regt.
26. "Coast Artillery Orders," *CAJ*, Vol. 83, No. 6 (November-December 1940), pp. 582-3.
27. Conn S. Rose Engleman, R. and Fairchild, B. (1964) pp. 47-8.
28. Ibid. Additional 16-inch and 6-inch batteries were authorized for the overseas bases.
29. Ibid.
30. Emanuel Raymond Lewis, *Seacoast Fortifications of the United States: An Introductory History* (Washington: Smithsonian Institution Press, 1970), p 115-118.
31. *NY-Phil*, pp.29-30.
32. Ibid., p. 185. Robert D. Zink, "Forts of Wherever #9, The Seacoast Defenses of Philadelphia, the Delaware River, and Delaware Bay," *CDSG News*, Vol. IV, No. 4 (November 1990), pp. 55-56.
33. *NY-Phil*, p. 184. "U.S. Guns to Canada," *CDSG News*, Vol. 2, No. 1 (January 1987), p.11 and Vol. 2, No. 2 (April 1987), p.17.
34. *NY-Phil*, pp. 31, 180-185.
35. *NY-Phil*, pp. 152, 174-176. *History of the Southeastern Sector of the Eastern Defense Command 1 March 1944 to 15 May 1945* (hereafter *SE Sector)*, Appendix II, pp.20-22.
36. C.W. Warrington, *A Mighty Fort Called Miles*, (Wilmington, Del: 1972), p. 15. Eric A. Pearson (Comp.), *Bits and Pieces on Fabulous Cape Henlopen*, (Lewes, Del.: 1993), pp.8-9.
37. *NY-Phil*, p.152.

38. Report of Inspection Trip by Lester L. Lessig, chief, Control Branch, Philadelphia Engineer District, 28 June, 1943. RCW, Fort Miles, Battery Smith (118), 1 May 1944.
39. RCW, Fort Miles, Battery Smith (118), 1 May 1944.
40. RCW, (Fire Control or Submarine Mine Structures), Fort Miles, Plotting Room, Battery Smith (118), 1 May 1944.
41. *NY-Phil*, p.177.
42. *NY-Phil*, pp.151-153.
43. Ibid., p. 156.
44. Ibid., p. 31.
45. Ibid., pp.27-28.
46. John C. Austin, "Second Coast Artillery District," *CAJ* Vol. 84, No. 5 (Sep-Oct 1941), p.505. *NY-Phil*, p.28.
47. *NY-Phil*, p.28. RCW, Fort Delaware, Searchlight No. 9, September 1, 1922, correction of February 28, 1939. "Fort DuPont, Delaware," New York Times, August 2, 1941, p.7.
48. General Miles had served as a 1st lieutenant in the 22nd Mass. Vol. Inf. during the Civil War, rising to major general of volunteers. He was appointed colonel of the 40th U.S. Infantry Regiment; again, became a general officer in 1880, this time in the regular army; and on April 5, 1890, was promoted to major general. General Miles was on active service, generally in the west, from 1869 until 1886. In 1895, he succeeded Major General John M. Schofield as commanding general of the army, and was the last officer to hold that title. In 1900, he was promoted to Lieutenant General. He ended a forty-two-year army career and retired on August 8, 1903. He died in Washington, D.C., on May 15, 1925. Francis B. Heitman, Historical Register and Dictionary of the United States Army from *Its Organization, September 29, 1789, to March 2, 1903*, Vol. I, pp. 708-709. "Fort is Named for Miles," *New York Times*, August 8, 1941, p. 9. "Forts Miles and Winslow," *CAJ*, Vol. 84, No. 6 (November-December 1941), p. 591-592.
49. "Fort Miles," *CAJ*, Vol. 84, No. 5 (September-October 1941), p. 491.
50. "Project at Fort Miles Halted by Strike," *New York Times*, August 27, 1941, p. 11. Report of Inspection Trip by Lester L. Lessig, chief, Control Branch, Philadelphia Engineer District, 28 June, 1943.
51. *NY-Phil*, p. 174.
52. Ibid., p. 175. "Fort DuPont," *New York Times*, December 2, 1941, p. 10.
53. William H.J. Manthorpe, Jr, *A Century of Service: The US Navy on Cape Henlopen, Lewes, Delaware: 1898-1996* (hereafter *Century of Service*), (Wilmington, DE, Cedar Tree Books, Ltd., 2014), p. 55-59.
54. Ibid, p. 60.
55. Ibid, p. 61.
56. "Fort DuPont," *New York Times*, September 27, 1941, p. 10.
57. "Fort DuPont," *New York Times*, October 25, 1941, p. 21.
58. Ibid, p. 21.
59. G.O. 11, AGO, October 14, 1941. *NY-Phil*, p. 177. "Army Honors Two of Its Dead," *New York Times*, November 18, 1941. p. 28.
60. *NY-Phil*, p. 151. War Department Seacoast Armament Board Report, November 1, 1946; in Casemate Museum, Fort Monroe. "Fort DuPont," *New York Times*, September 27, 1941, p.10.
61. "Fort Miles Bars Civilians," *New York Times*, December 2, 1941, p.10.
62. *NY-Phil*, p. 175.
63. Ibid., p.149.
64. *EDC*, p.201
65. *NY-Phil*, pp. 151-153. RCW, Four 155 mm Guns, Cape May, N.J., 1 January 1945.
66. "Heavy Firing off Coast," *New York Times* December 23, 1941. p.18
67. *NY-Phil*, pp. 92-94, 150, 153-154. Schwarzkopf, former head of the N.J. State Police, had led the investigation of the Lindberg kidnaping. He was called to active duty when the 113th Infantry of the N.J.N.G. was federalized in 1940. 55. Ibid., p. 147-154,
68. Ibid., p. 154-155.
69. Ibid., p. 154-156.
70. Robert D. Zink, "Controlled Submarine Mining in the United States," *CDSG Journal*, Vol. 9, No. 4 (November 1995) (hereafter Zink, "Mining"), p.45.
71. S.J. Leahy, "New York City and Delaware River/Bay," *CDSG News*, Vol. 2, No.2 (April 1987), pp.5- 6.
72. *SE Sector*, Appendix II, p.17.
73. USAMP *General John M. Schofield* was one of the oldest mine planters still in service, having been commissioned in 1909. The 165-foot, 601-ton, vessel was one of two built by the New York Shipbuilding Company of Camden, N.J. K.L. Waters, "The Army Mine Planter Service," *Warship International*, No. 4, 1985, pp. 401-402.
74. *NY-Phil*, p. 167.
75. Zink, "Mining", p. 45. *SE Sector*, Appendix II, p. 17.
76. *NY-Phil*, p. 155.
77. Zink, "Mining", p. 45. *NY-Phil*, p. 151, 155. War Department Seacoast Armament Board Report, November 1, 1946.
78. *NY-Phil*, p. 90, 167.
79. Ibid., p. 156. RCW, Fort Miles, Searchlight Shelter, 1 May 1944. RCW, Fort Saulsbury, Searchlight Shelter, 1 May 1944. RCW, Fort Saulsbury, F.C. Station Location No. 8, 1 May 1944. RCWs, Fort Saulsbury, F.C. Station Location No. 9, 10, 11, and 12, all 1 May 1944. *Coast Artillery: A Complete Manual of Technique and Materiel*, (Harrisburg, PA: 1942), pp. 622-623, 629-634.
80. RCW, Fort Miles, Battery Herring, 1 May 1944.
81. Harbor Defense Project Annex A, Seacoast Guns, 1 July 1945, Harbor Defense Projects for Harbor Defenses included in the Philadelphia-Delaware River Sub-Sector, EDC.
82. *NY-Phil*, p. 157-158.
83. Ibid., p. 156-157, 178, 185. RCW, Fort Miles, Battery No. 5, 1 May 1944.
84. *Century of Service*, pp. 59.
85. Ibid, p. 60.
86. Ibid, p. 61-62.
87. William H.J. Manthorpe, Jr, *A Century of Service: The US Navy on Cape Henlopen, Lewes, Delaware: 1898-1996* website "Submarines at the Cape: Friend and Foe", http://navyatcapehenlopen.info/plussubmarinesatcape.html .
88. Michael Gannon, *Operation Drumbeat: The Dramatic True Story of Germany's First U-Boat Attacks Along the American Coast in World War II* (hereafter *Drumbeat*), (Annapolis, MD, Naval Institute Press., 2009), p.24.
89. Ferguson, Arthur B. (April 1945). "*The Antisubmarine Command, USAF Historical Study No. 107*" (Washington DC, Assistant Chief of Air Staff, Intelligence Historical Division, 1945), pp.189-192
90. *Century of Service*, webpage p. 8.
91. Ibid, webpage p.9.
92. Ibid, webpage p.10.
93. Ibid, webpage p.12.
94. Ibid, webpage p.13.

95. Ibid, webpage p.15.
96. Harbor Defense Project Annex A, Seacoast Guns, 1 July 1945, Harbor Defense Projects for Harbor Defenses included in the Philadelphia-Delaware River Sub-Sector, EDC. p.182
97. RCW, Fort Miles, Battery Hunter, 1 May 1944.
98. G.O. 46, AGO, 17 September 1942.
99. *SE Sector*, Appendix II, p.18. *NY-Phil*, p.160.
100. *NY-Phil*, p. 161, 162-163.
101. Charles L. Combes, "The Railway Artillery is Ready to Roll," *CAJ*, Vol. 85, No. 6 (November-December 1942), pp.6-10. *NY-Phil*, pp.159-160. RCW, Fort Miles, Temporary Battery 8-inch Railway Mount, 1 August 1944.
102. *NY-Phil*, pp. 80-86, 160. RCW, Fort Miles, Temporary Battery 8-inch Railway Mount, 1 August 1944.
103. *NY-Phil*, p. 24, 148, 158-159, 180-181.
104. *NY-Phil*, p. 180-181, Harbor Defense Project Annex A, Seacoast Guns, 1 July 1945, Harbor Defense Projects for Harbor Defenses included in the Philadelphia-Delaware River Sub-Sector.
105. RCW, Fort Miles, Battery Smith, 1 May 1944.
106. History of Development of 16" Gun Materiel Used for Seacoast Defenses, (U.S. Army Ordnance Department, undated) p. 13-22, 24-25, 44-45.
107. RCW, Fort Miles, Battery 519, 1 May 1944.
108. RCW, Fort Miles, Battery 519, 1 May 1944.
109. RCW, Fort Miles, Plotting Room, Battery 519, 1 May 1944.
110. *NY-Phil*, pp. 160, 170-171.
111. Ibid., pl. 16, following p.170.
112. Ibid., pp.41-43, 79-80.
113. Ibid., pp.88-92.
114. Ibid., pp.120-121.
115. Elizabeth G. R. Ross, *FORT MILES: ADAPTATION, RESILIENCE AND A MILITARY LEGACY*, A thesis submitted to the Faculty of the University of Delaware – Spring 2002 (hereafter *Military Legacy*), Chapter 5, p.2
116. *Military Legacy*, Chapter 5, p.3
117. Ibid, Chapter 5, p.4
118. Ibid, Chapter 5, p.6
119. Ibid, Chapter 5, pp.7-8
120. Ibid, Chapter 5, p.8
121. Ibid, Chapter 5, p.10
122. Ibid., pp.168, 170, 171, 179. RCW, Fort Miles, Battery 5A, 1 January 1944. RCW, Fort Miles, Battery 5B, 1 January 1944.
123. *NY-Phil*, pp.158,168, 179. RCW, Fort Miles, Battery 5A, 1 May 1944, RCW, Fort Miles, Battery 5B, 1 May 1944.
124. *NY-Phil*, p.179. RCW, Cape May, New Jersey, Battery 7, 1 May 1944.
125. *NY-Phil*, pp.171, 182-185.
126. Ibid., p.171. The engineers did not redesignate the plotting and switchboard room for the once planned second 16-inch casemated battery when it was cancelled and redesignated Battery 519; the command post structure retained its original numerical designation.
127. RCW, Fort Miles, Harbor Defense Command Post, 1 May 1944.
128. RCW, Fort Miles, F.C. Tower No. 12, Harbor Defense Observation Post, 1 May 1944. RCW, Fort Miles, SCR-582.
129. *NY-Phil*, p.176.
130. *Ibid.*, p.176. RCW, Fort Miles, Location no. 5, Fire Control Tower No. 9, 1 May 1944.
131. RCW, Fort Miles, Battery Hunter, 1 May 1944
132. *NY-Phil*, p. 148. RCW, Fort Miles, Battery Herring, 1 May 1944. RCW, Fort Miles, Battery Hunter, 1 May 1944, RCW, Cape May, N.J., Battery 223, 4 May 1944. Robert D. Zink, "The Six-inch Part of the Modernization Program of 1940," *CDSG Journal*, Vol. 8, No. 2, (May 1994), p.24.
133. *EDC*, p. 9. RCWs, Fort Miles, SCR-296 Nos. 1, 2, 3, 4, and 5, all 1 August 1944.
134. *SE Sector*, p.29. Basic Project for The Harbor Defenses of The Delaware 1945, Annex A (Armament).
135. *NY-Phil*, p.179.
136. *NY-Phil*, p.173.
137. *EDC*, p.28.
138. *SE Sector*, p.29.
139. RCW, Cape May, N.J., Battery No. 26, Four 155 mm Mobile Carriages, January 1, 1945. RCW, Fort Miles, Battery No. 22, Four 155 mm Mobile Carriages, May 1, 1944.
140. *EDC*, p.28. Data Sheet, 21st C.A.(HD) Bn.(Sep).
141. *EDC* p.29.
142. *SE Sector*, p.50.
143. Data Sheet, HQ and HQ Battery, HD of the Delaware.
144. "Navy Ends Control at 13 Ports," *New York Times*, August 19, 1945, p.23.
145. Data Sheet, HD of the Delaware.
146. Coast Defense Study Group, "*Forts, Military Reservations and Batteries 1794-1945*", CDSG website 2020, webpage (https://cdsg.org/fort-and-battery-list/)
147. Report of the War Department Seacoast Defense Armament Board, November 1, 1946.
148. Warrington, C. W. *Delaware's Coastal Defenses: Fort Saulsbury & a Mighty Fort Called Miles*. Delaware Heritage Press, 2003. p. 98.
149. Hamilton, Michael A, and George W Contant. "Fort Miles: The Cold War in Miniature." *Outdoor Delaware*, 2011, p. 20.
150. *Century of Service*, pp. 76-79.
151. *Delaware's Coastal Defenses*, p. 99.
152. *Century of Service*, p.83.
153. Hardy-Heck-Moore & Associates, Inc. "Naval Reserve Facility, Lewes, Delaware." Austin, Tx, 27 Sept. 1996. p.10.
154. *The Cold War in Miniature*, p.21.
155. *Century of Service*, pp.84-85.
156. Whitman, E. SOSUS: The "secret weapon" of undersea surveillance. *Undersea Warfare*. 2005.. p.89.
157. *Century of Service*, p. 94-97.
158. *The Cold War in Miniature*, p. 21.
159. *Century of Service*, p. 87-91, 119.
160. *Naval Reserve Facility, Lewes, Delaware*, p. 12.
161. *Century of Service*, p. 128-129.
162. Fort Miles Historical Association, Lewes, Delaware. Website: *http://.fortmilesha.org/*

Appendix

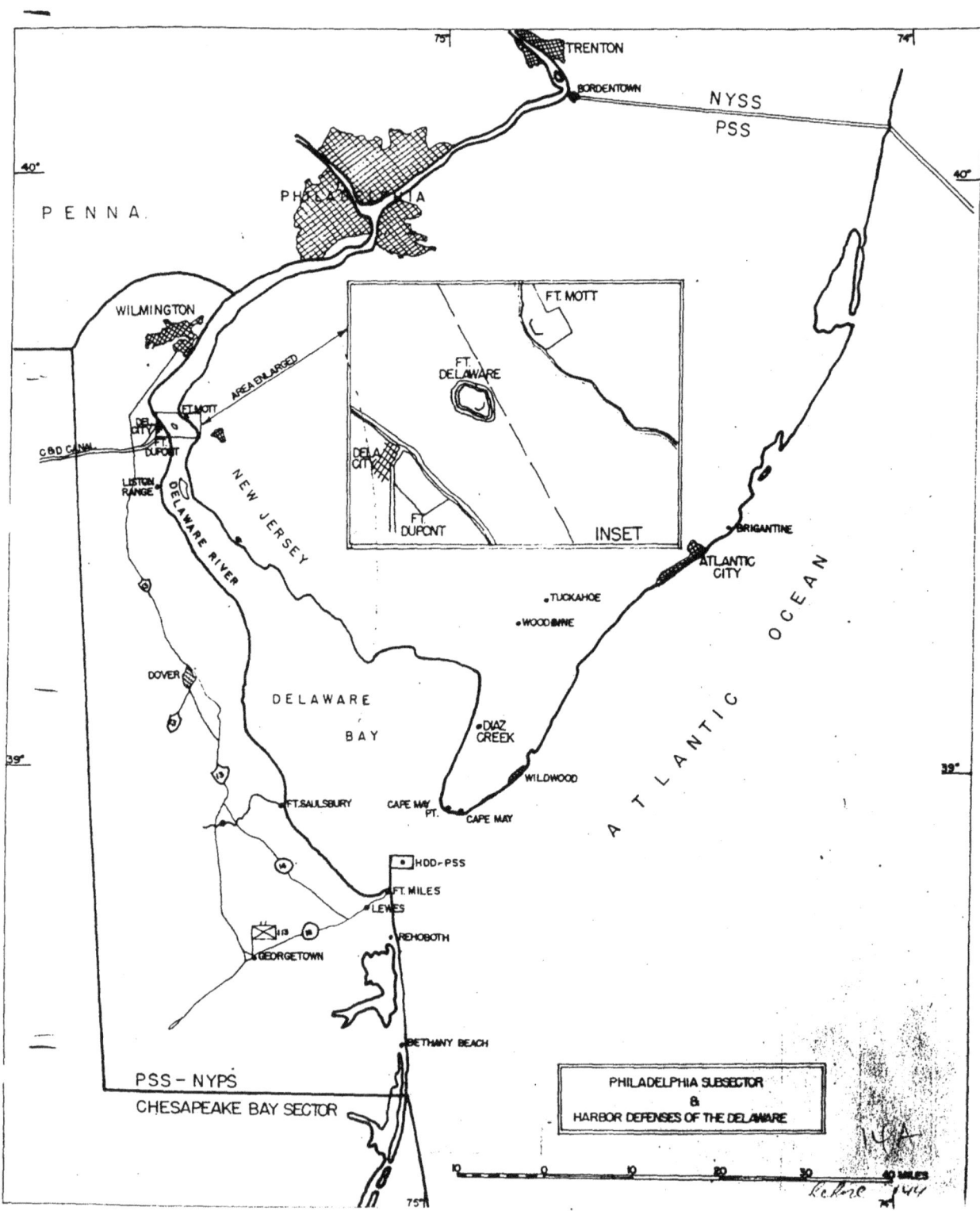

Figure 3. The Philadelphia Subsector and the Harbor Defenses of the Delaware during World War Two. (NY-Phil History 1945)

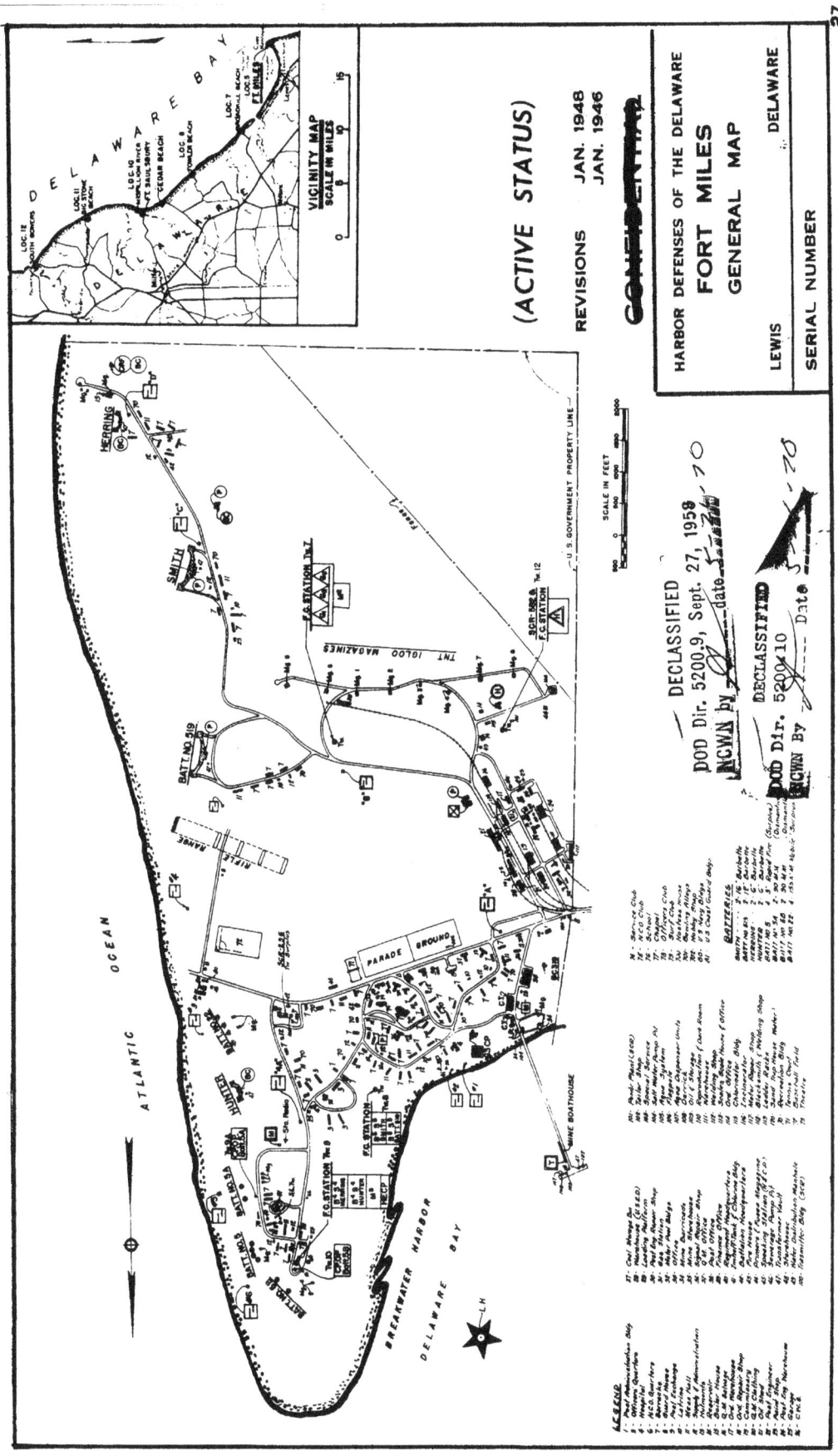

Figure 56. The general map of Fort Miles, DE showing the numerous structures built between 1941 and 1944 to support the fort's coast defense role as only a few buildings existed at Cape Henlopen before Fort Miles was established and by the date of this map in 1948 there is a city of building. (RG177 NARA 1948)

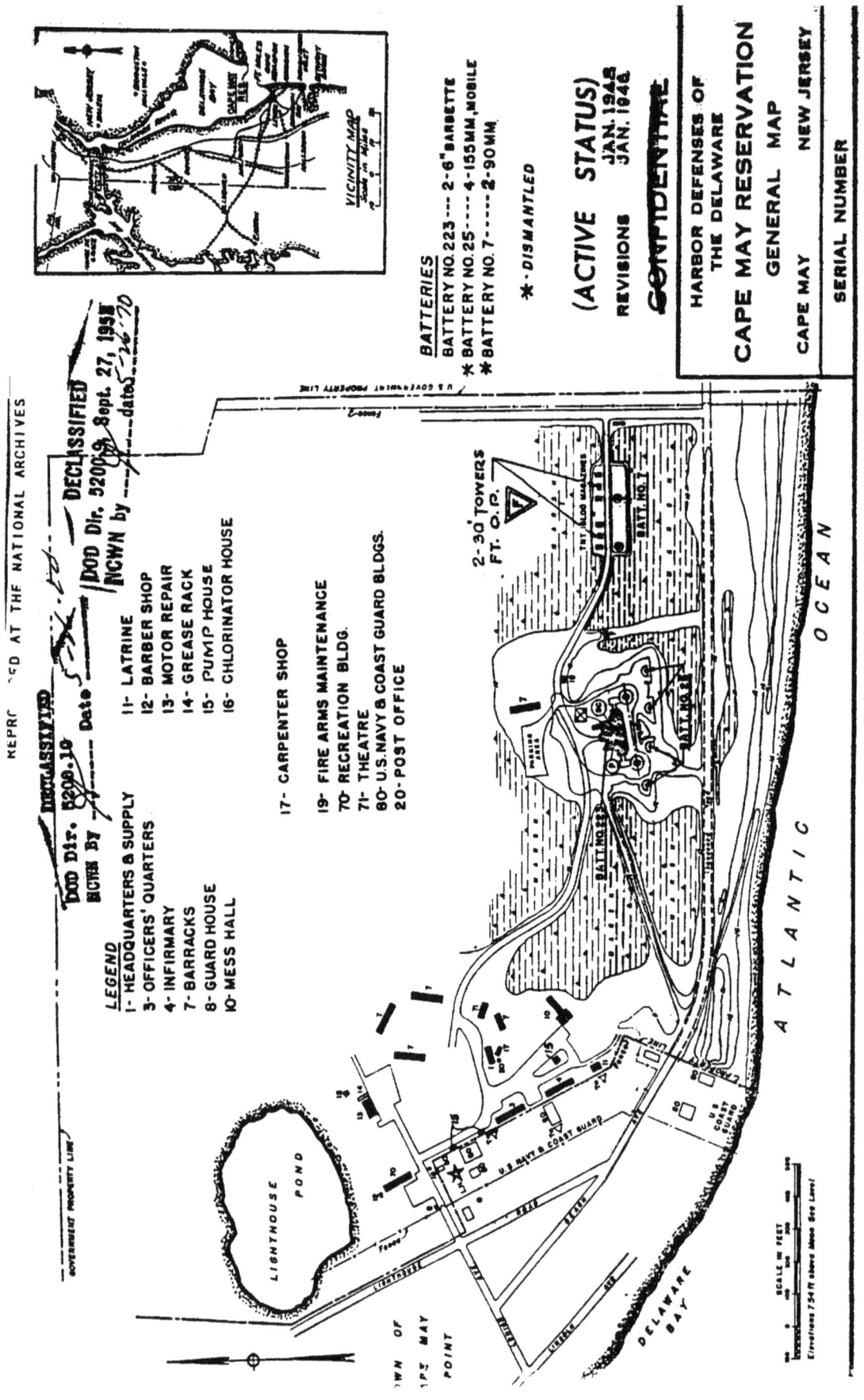

Figure 61. General map of defenses of Cape May, NJ with Battery #223 (two 6-inch shielded guns), Battery #7 (four 90mm guns), and the temporary Battery #25 (four 155mm GPF on Panama mounts) in January 1948. (RG177 NARA 1948)

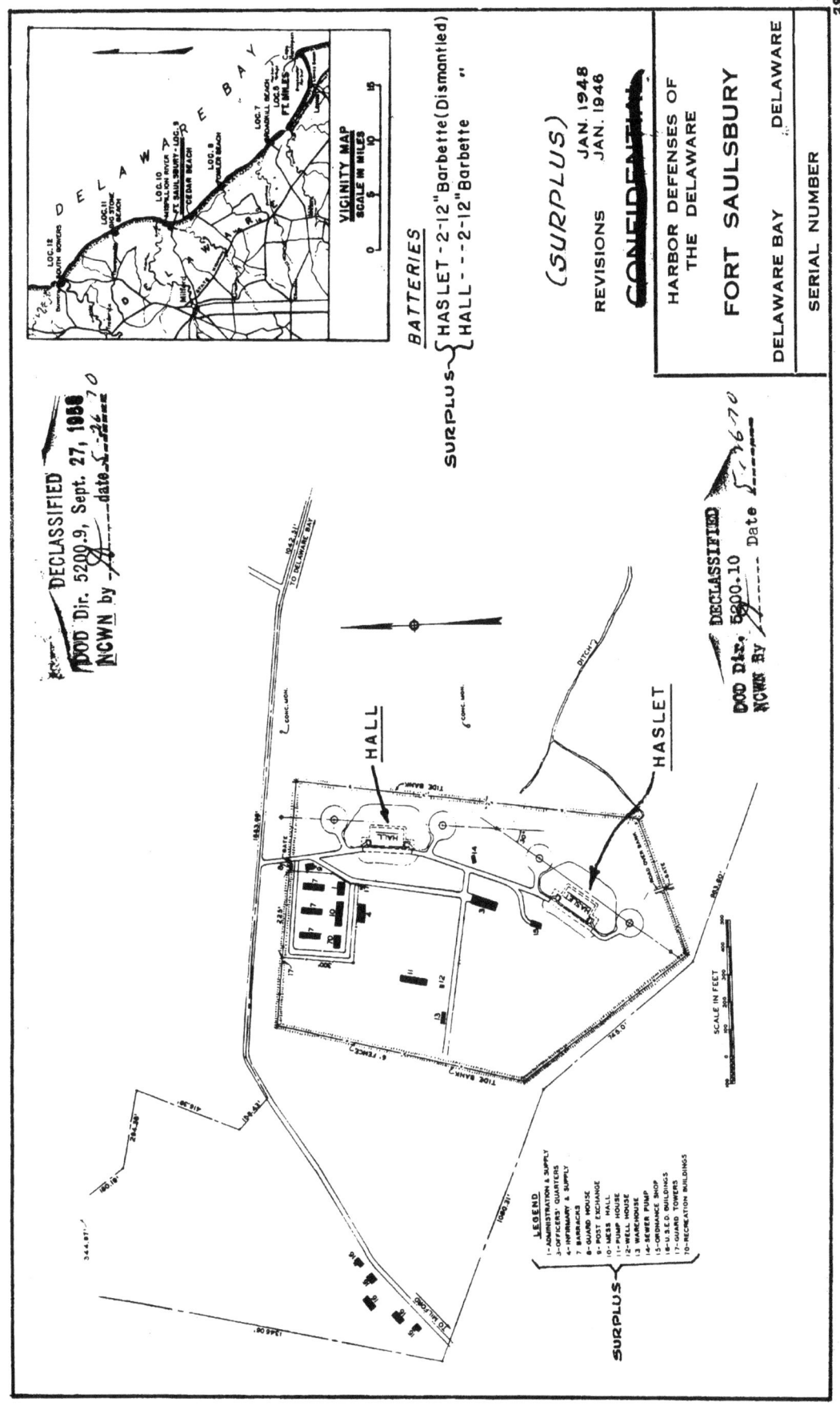

Figure 48. Final site plan for Fort Saulsbury in January 1948 shows the new WW2 support buildings while guns of Batteries Hall and Haslet have been dismantled and fort is now surplus of US Army needs. (NARA 1948)

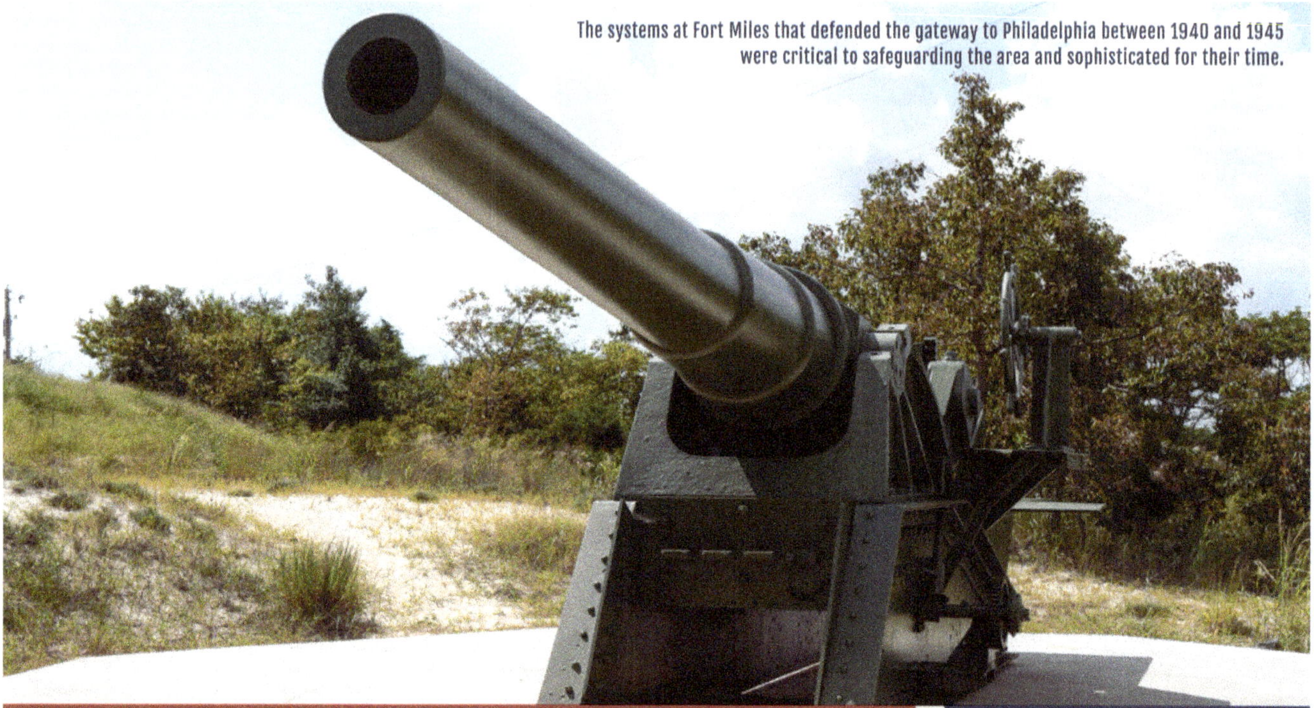

The systems at Fort Miles that defended the gateway to Philadelphia between 1940 and 1945 were critical to safeguarding the area and sophisticated for their time.

Be part of a history-making mission!

FORT MILES HISTORICAL ASSOCIATION

THE FORT MILES HISTORICAL ASSOCIATION is working tirelessly to develop the Fort Miles Museum into a world class cultural resource. The Museum, which is located in a wartime coastal fortification, provides a unique experience for visitors to coastal Delaware. FMHA's mission is to honor Delaware's fallen veterans and educate visitors about the role of Fort Miles in World War II and the Cold War that followed.

You can be part of this exciting mission. Become a member today.

Learn more at fortmilesha.org

Fort Miles Historical Association, PO Box 52, Lewes, Delaware 19958

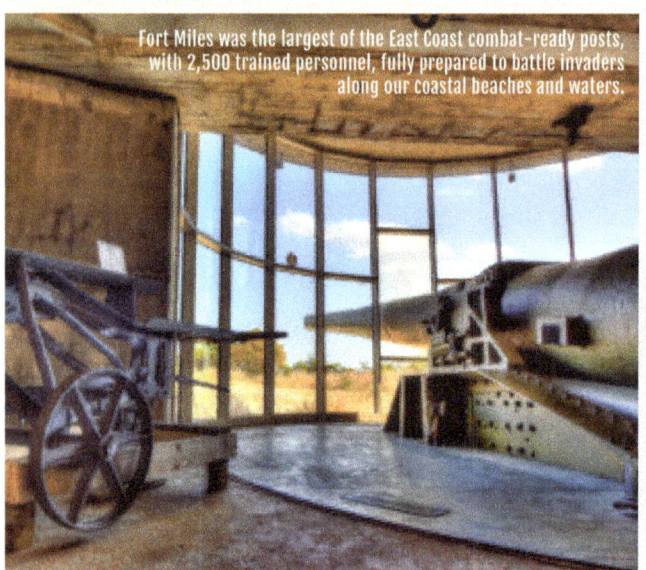

Fort Miles was the largest of the East Coast combat-ready posts, with 2,500 trained personnel, fully prepared to battle invaders along our coastal beaches and waters.

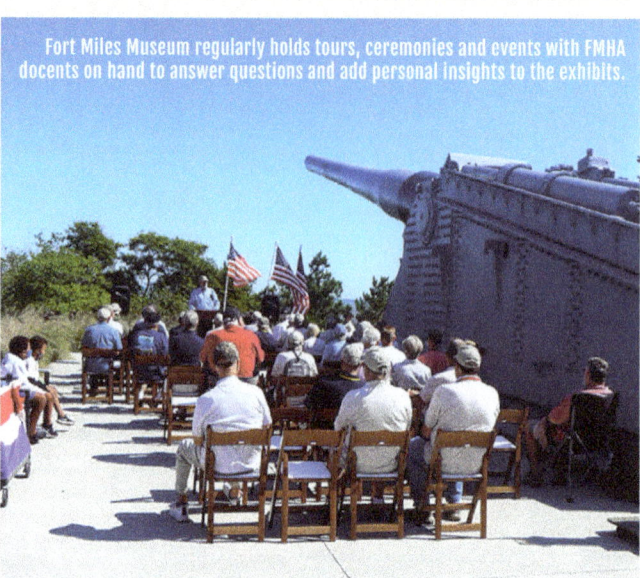

Fort Miles Museum regularly holds tours, ceremonies and events with FMHA docents on hand to answer questions and add personal insights to the exhibits.

Fifteen fire control towers were used to sight enemy surface ships and to assist sighting the minefield entrance to the Delaware Bay.

Fort Miles | an American treasure

FORT MILES MUSEUM

The Fort Miles Museum brings the historic WW II era to life.

Fort Miles was constructed during World War II as a vital coastal defense installation to deter potential German attacks on key U.S. infrastructure and port cities, such as Philadelphia and Wilmington. Later, it was home to the Top-Secret US Navy SOSUS program.

The Fort Miles Museum, located underground in Battery 519, features exhibits relating to coastal defense operations as well as portraying daily life for the over 2,000 soldiers and civilians stationed at Fort Miles during World War II.

The Fort Miles Museum continues to expand its World War II exhibit offerings and community outreach initiatives to tell the stories of Delaware residents and veterans.

Support the Cause. **Donate today.**

Visit Battery 519, the Artillery Park, and Historical Area located in Cape Henlopen State Park in Lewes, DE.

fortmilesha.org
Fort Miles Historical Association
PO Box 52, Lewes, Delaware 19958

The Plotting Room features a plotting table used to calculate the required trajectory of fire in order to strike a ship entering the Delaware Bay.

Visitors to Fort Miles Museum can enjoy a breathtaking view of the Delaware Bay.

The Coast Defense Study Group

The **Coast Defense Study Group, Inc.** (CDSG) is a tax-exempt corporation dedicated to study of seacoast fortifications. CDSG's purpose is to promote and encourage the study of coastal defenses, primarily but not exclusively those of the United States of America. The study of coast defenses and fortifications includes their history, architecture, technology, strategic and tactical employment and evolution. The primary goals of the CDSG are the following:

- Educational study of coast defenses
- Technical research and documentation of coast defenses
- Preservation of coast defense sites, equipment and records for current and future generations
- Accurate coast defense site interpretations
- Assistance to groups interested in preservation and interpretation of coast defense sites
- Charitable activities which promote the goals of the CDSG

The CDSG was officially founded in 1985 at Fort Monroe, Virginia and incorporated in 1993. The first "St. Babs" conference was held in 1978 to visit the harbor defenses of New York. Subsequent annual CDSG conferences have been held at all the major harbor defenses of the continental Unites States. Several special overseas tours have also been organized. Membership is open to any person or organization interested in the study or history of the coast defenses and fortifications. Membership in the CDSG will allow you to attend the annual conference, special tours and receive the CDSG's quarterly journal and newsletter. For more information on the CDSG, please visit the CDSG website at www.cdsg.org or contact us at 24624 W. 96th Street, Lenexa, KS 66227-7285 USA, Attn: Quentin Schillare, Membership.

The **CDSG Fund** supports the efforts of the Coast Defense Study Group by raising funds for preservation and interpretation of American seacoast defenses. The CDSG Fund is seeking donations for projects supporting its goals. Donations are tax-deductible for federal tax purposes as the CDSG is a 501(c)(3) organization, and 100% of your gift will go to project grants. Major contributions are acknowledged annually. The Fund is always seeking proposals for the monetary support of preservation and interpretation projects at former coast defense sites and museums. A one-page proposal briefly describing the site, the organization doing the work, and the proposed work or outcome should be sent to the address below. Successful proposals are usually distinct projects rather than general requests for donations. Upon conclusion of a project a short report suitable for publication in the CDSG Newsletter is requested. The trustees shall review such requests and pass their recommendation onto the CDSG Board of Directors for approval. Send donations and grant requests to: CDSG Fund c/o Terry McGovern 1700 Oak Lane McLean, VA 22101-3326 USA or use your credit card via PayPal on the www.cdsg.org website.

The CDSG ePress

The CDSG Digital Library

The CDSG has digitized an extensive set of historic manuals, reports, records and documents on the harbor defenses of the United States Army. The documents have been broken down to a general collection of manuals and reports (CDSG Documents collection) and also the records of the various harbor defenses so you can order sets of records by coast/harbor defense or get the complete collection (CDSG Harbor Defense Collection). The CDSG provides its back issues of the CDSG Publications in electronic format. **Back Issues CDSG Publications DVD** costs $55 and annual update (with original DVD insert) $10. The **CDSG Documents USB drive** covers a range of historical material related to seacoast defenses -- most are from the National Archives. Included are the annual reports of the chief of coast artillery and chief of engineers; several board proceedings and reports; army directories; text books; tables of organization and equipment; WWII command histories; drill, field, training manuals and regulations; ordnance department documents; ordnance tables and compilations; and the ordnance gun and carriage cards. This UBS drive of CDSG Documents costs $50. Order the CDSG ePress items directly from Mark Berhow at PO Box 6124, Peoria, IL 61601 USA or at berhowma@cdsg.org and through www.cdsg.org.

Documents related to specific harbor defenses. These PDF documents form the basis of the conference and special tour handouts that have been held at these locations. These collections are available as PDF files on DVD. They include RCBs/RCWs; maps; annexes to defense projects; CD engineer notebooks; quartermaster building records; and aerial photos taken by the signal corps 1920-40. Please consult www.cdsg.org for more details. Contact Mark Berhow by post/email if you are interested in ordering only specific titles.

The CDSG Press

Prices Include Domestic/International Postage $US currency only (cash, check, money order), allow 6-8 weeks for delivery - CDSG Books and CDSD Gear ($ domestic / $ International)

Notes on Seacoast Fortification Construction by Col. Eben E. Winslow, 1920, 428 pp. 1994 reprint HC with drawings $45/$60
Seacoast Artillery Weapons Technical Manual (TM) 9-210 by U.S. War Dept. 1944, 202 pp. 1995 reprint HC $25/$35
The Service of Coast Artillery by F. Hines & F. Ward, 1910, 736 pp. 1997 reprint HC $40/$60
Permanent Fortifications & Sea-Coast Defences by U.S. Congress, 1862, 544 pp. 1998 reprint HC $30/$45
American Coast Artillery Matériel Ordnance Dept. Doc#2042 by U.S. War Dept., 1922, 528 pp., 2001 reprint HC $45/$65
American Seacoast Defenses: A Reference Guide (3rd Edition) by Mark A. Berhow, (2015) 732 pp. HC $45/$80
The Endicott-Taft Reports, reprint of original reports of 1886 and 1905 by U.S. Congress, 525 pp. 2007 reprint HC $45/$80
Artillerists and Engineers: The Beginnings of US Fortifications 1794-1815 by Col. Wade, U.S. Army. PB, 226 pp. $25/$40

CDSG Logo Hats each $20.00 domestic and $25.00 foreign.   **CDSG Logo Patches** each $ 4.00 domestic & foreign.
CDSG T-Shirts (XXXL, XXL, XL, L; Red, Khaki, Navy, Black) $18.00 Domestic and $26.00 Foreign.
Send Order to: CDSG Press Attn: Terry McGovern 1700 Oak Lane, McLean, VA 22101-3326 or order via www.cdsg.org

McGovern Publishing Presents:

McGovern Publishing is comprised of two divisions: Redoubt Press (military titles) and Three Sisters Press (Rebecca, Rachel, and Alana) offering a range of subjects. McGovern Publishing is interested in new titles, especially those dealing with fortifications, please contact Terry McGovern at 703/538-5403 or at tcmcgovern@att.net if you have a title that you are seeking to have published. Visit our website at www.mcgovernpublising.com, or post to 1700 Oak Lane, McLean, Virginia 22101-3326 USA

Redoubt Press
A Division of McGovern Publishing

A Division of McGovern Publishing

Redoubt Press titles:

American Defenses of the Panama Canal
By Terrance McGovern.

The end of 20th Century has brought to a close the American involvement in the Panama Canal, a tremendous technological achievement that was started in the early days of the century. The Panama Canal took over 10 years to complete and involved the expenditure of millions of dollars and thousands of workers lives. The importance of the Panama Canal to commerce and naval forces resulted in fortifications that match its size and cost. While the history of the Panama Canal has been recorded in numerous books and articles, the history of its defenses has not. This book hopes to remedy this situation by exploring the thirty-five years of fortification construction and use in the Panama Canal Zone.

The first part of this book discusses how the various fortification boards decided, with the help of Congress, on what defenses were needed and how evolving threats resulted in changes to these defenses. The second part of the book describes in detail the construction, service life, and current status of each major battery. Assisting this narrative are many current and period photographs, along with engineering drawings and site maps detailing the construction of these impressive fortifications. All serious students of seacoast defenses and the Panama Canal should have a copy.

This title was originally published in U.K based Fortress Study Group annual journal, *FORT: The International Journal of Fortification and Military Architecture* (Volume 26: 1998). This 116-page book is softbound using the same high-quality paper and printing process as *FORT*. As the most detailed and illustrated article to ever be produced on these defenses, the book has 45 maps and drawings, 43 historical B&W photographs, and 40 color photographs (taken in 1993 and 1999 by the author). This book is offered at a price of $55 plus $5 for domestic shipping or $10 for foreign shipping.

The Concrete Battleship
Fort Drum, El Fraile Island, Manila Bay
by Francis J. Allen

Fort Drum on El Fraile Island in the Philippines is unique in the development of United States coastal fortifications. Fort Drum is part of a chain of forts built across the entrance of Manila Bay to defend the Bay from naval attack. The construction of Fort Drum began in 1909 by reducing tiny El Fraile Island to the low water mark. Over the next ten years a multi-deck concrete island was built to mount two twin 14-inch guns in superimposed Army designed armored turrets. The completed work rises 40 feet above sea level, it is 350 feet long and 144 across at its widest point. The exterior walls are up to 28 feet thick and the top deck attains a thickness of 20 feet of re-enforced concrete. The interior of the fort held a large engine room, powder and shell magazines, a mining casemate, storerooms and tankage, a accommodations for 300 personnel. The design of the fort followed a naval pattern with turrets, a cage mast, and secondary armament in side casemates. Due to these characteristics, Fort Drum became known as the "Concrete Battleship."

When completed in 1918, Fort Drum was the most powerful defense work in Manila Bay, but the advances in military technology during World War I already began to make the fort obsolete. The post-World War I reduction in military spending, the re-strictions of the Washington Naval Treaty of 1922, and economic depression of the 1930s resulted in Fort Drum being quickly reduced to caretaker sta-tus until the coming of World War II. Fort Drum became an important weapon during the Japanese siege of Corregidor and the other island forts during 1942 but only play a minor role during the American retaking of these islands in 1945. The battles of World War II would transform Fort Drum from an American-manned, fully operating fort to a burned-out hulk inhabited by lifeless Japanese sailors. This revised and enlarged 64-page softbound volume tells the story of The Concrete Battleship in words, diagrams, and photographs from its inception to the present day. Redoubt Press is pleased to offer this book for $20 plus $5 for domestic shipping or $10 for foreign shipping.

A Legacy in Brick and Stone – 2nd Edition
American Coastal Defense Forts of the Third System
1816-1867
By John R. Weaver II

The definitive history of the American Third System of Fortifications that defended our coastline for more than half of century, these architectural wonders were built from 1816 through 1867 from Maine through the Florida Keys to New Orleans, with two forts in San Francisco Bay. Most of these 42 masonry forts still stand guard along our shores, and open to the public. *A Legacy in Brick and Stone* provides the background of these famous Civil War forts – why they were built where they are, who built them, and how they functioned – as well as descriptions of each of the forts.
This revised and expanded edition has grown to 340 pages with 400 new photographs and drawings. John Weaver II, a nationally known expert on masonry coastal fortifications, has invested over 30 years of research into *A Legacy in Brick and Stone* to produce the only a full treatment of the magnificent American Third System.
  The book begins with a study of the history of the Coastal Fortifications Board, which developed and implemented this massive defense project. It then details the art of fortification of that period and describes the particular architecture components that were key to their design. A description of the development of the system over its 50-year life is followed by an analysis of how well certain forts held up under attack during the American Civil War. Approximately two-thirds of this volume is dedicated to a fort-by-fort description of the system. The overall defense scheme for each harbor is discussed, then each fort in that harbor is analyzed. The author uses unique photographs and drawings to answer many of the questions about of these forts that today's visitors ask. The book also documents the current status of these historic forts, including information about how to visit these forts today.
  Redoubt Press is pleased to offer a deluxe hardcover edition with color illustrations $65 plus a $5 fee for domestic shipping and $10 for foreign shipping. Paperback edition with black and white illustrations $40 plus a $5 fee for domestic shipping and $10 for foreign shipping.

Pacific Rampart
A History of Corregidor and Harbor Defenses of Manila and Subic Bay
by Glen M. Williford

This is the definitive history of the forts and harbor defenses of Manila and Subic Bay in the Philippines. The heavily illustrated work tells the story of these fortified islands, including the famed island of Corregidor, from the fortifications built by the Spanish in 1898 to the Japanese defenses in 1945. Drawing on years of research from the National Archives and many primary sources, this book describes the various proposed defensive plans and the fortifications actually built by the US Army between 1904 and 1942. It chronicles the heroic defense and desperate fighting early in the Second World War that led to the famous surrender of these defenses, as well as the intense combat in early 1945 when these fortified islands were retaken from the Japanese. This book is simultaneously a unit history (primarily of the Coast Artillery units stationed on the islands over almost 50 years), a technical history (the impressive weapon systems and their supporting garrison infrastructure), and a combat history (taking and then retaking of the fortress during the Second World War). This 470-page, hardcover book is thoroughly supported with endnotes, a bibliography, six appendixes, and over 340 illustrations (black and white photographs, maps, and plans for many of the islands' structures). The author, Glen M. Williford, has invested over 30 years of research into Pacific Rampart, making it the primary source on these historical defenses and a must for any serious student of these fortifications and the fascinating history of Corregidor and Manila and Subic Bays.
  Redoubt Press is pleased to offer this 470-page hardcover edition for $50 plus a $5 fee for domestic shipping and $10 for foreign shipping.

Pacific Fortress
The Harbor Defenses of Oahu, Hawaii
by Glen M. Williford.

Forthcoming title containing the definitive history of the forts and harbor defenses of Oahu, Hawaii from 1898 to 1950. The heavily illustrated work will tell the story of the strongly defended island, including the famous raid on Pearl Harbor on December 7th, 1942.

Three Sisters Press titles:

The Chesapeake Bay at War!
The Coastal Defenses of the Chesapeake Bay During World War II
by Terrance McGovern

The defense of America's seacoast has been one of key concerns since the earliest years of the Republic. American coastal defenses steadily evolved through the age of muzzle loading cannon, ever larger breech loading weapons, and finally to the culmination of large, long-range guns capable of targeting the largest and most heavily armed warships of their age. By the end of World War II, the United States had some of the strongest defenses in the world. Given the importance of the U.S. naval bases around Norfolk, Virginia and the shipyards of Hampton Roads, the seacoast defenses protecting Chesapeake Bay contained the largest collection of firepower in the continental United States as they reached their apex during World War II.

This book tells the story of preparing the coastal defenses of the Chesapeake Bay for the coming of World War II and their operations from 1941 to 1945. Over a hundred rare black & white U.S. Army photographs and plans help document our nation's extensive efforts to defend against naval attacks and raids from Nazi Germany. A collection of over 50 recent color aerial photographs are included allowing the reader to survey the surviving elements of these mighty defenses. A product of extensive research this book brings together rare images and the little-known military history of the Chesapeake Bay for the first time. Three Sisters Press (Rebecca, Rachel, and Alana) is pleased to offer this 70-page softbound book for $20 plus $5 for domestic shipping or $10 for foreign shipping.

The Delaware Bay at War!
The Coastal Defenses of the Delaware Bay During World War II
by Terrance McGovern

The defense of America's seacoast has been one of the key concerns since the earliest years of the Republic. American coast defense steadily evolved through the age of muzzle loading cannon, ever larger breech loading weapons, and finally to the culmination in large, long-range guns capable of targeting the largest and most heavily armed warships of their age. By the end of World War II, the United States had some of the strongest defenses in the world. Given the importance of the military-industrial complex along the banks of the Delaware River, including the large Philadelphia Naval Shipyard, the seacoast defenses protecting Delaware Bay had declined dramatically since the turn of the century resulting in a whole program of modern coast artillery batteries and other defenses to be constructed starting in the late 1930's and reaching their reached their apex during the middle of World War II.

This book tells the story of preparing the coastal defenses of the Delaware Bay for the coming of World War II and their operations from 1941 to 1945. Over a hundred rare black & white U.S. Army photographs and plans help document our nation's extensive efforts to defend against naval at- tacks and raids from Nazi Germany. A collection of recent color aerial photography is also included allowing the reader to survey the surviving elements of these generally unknown defenses. A product of extensive research, this book brings together for the first-time rare images and the little-known military history of the Delaware Bay. Three Sisters Press (Rebecca, Rachel, and Alana) is pleased to offer this 65-page softbound book for $30 plus $5 for domestic shipping or $10 for foreign shipping.

Seacoast Cannon Coloring Book
By Brian B. Chin

The big seacoast guns thunder once again. From the decorated brass cannon of Spanish New World, to the powerful iron guns in the American Civil War, to the complex disappearing guns of the turn of the century, to the huge guns of steel and armor in World War II, this detailed picture book shows how coast artillery once protected American shores from enemy attack. This second edition brings to life the coast defenses that today remain as empty forts and batteries with our national, state, and local parks. The author, Brian B. Chin, is a well-known author and artist that has produce several works on military fortifications. This coloring book allow a new generation to learn the history of American seacoast defenses while having fun coloring in this 52-page softbound book. It is hoped that book will help generate interest in preserving and interpreting these historic sites for future generations.

Three Sister Press (Rebecca, Rachel, and Alana) are pleased to offer this coloring book for $10 plus $5 for domestic shipping or $10 for foreign shipping.

www.ingramcontent.com/pod-product-compliance
Lightning Source LLC
Chambersburg PA
CBHW061112070526
44583CB00027B/3271